Capitals of Punk

Tyler Sonnichsen

Capitals of Punk

DC, Paris, and Circulation in the
Urban Underground

Tyler Sonnichsen
University of Tennessee at Knoxville
Knoxville, TN, USA

ISBN 978-981-15-2673-2   ISBN 978-981-13-5968-2   (eBook)
https://doi.org/10.1007/978-981-13-5968-2

Library of Congress Control Number: 2019934468

Cover images: Capitol Hill © Luis Prado; Eiffel Tower © Pedro Santos; Arc de Triomphe © Iconika: White House © Leif Michelsen.
Cover design: Fatima Jamadar

This Palgrave Macmillan imprint is published by the registered company Springer Nature Singapore Pte Ltd.
The registered company address is: 152 Beach Road, #21-01/04 Gateway East, Singapore 189721, Singapore

Ian MacKaye, Arnaud Gabelli, and Philippe Roizès on the porch at Dischord House in Arlington, Virginia, Summer 1987. (Photo courtesy of Phillippe Roizès)

Banned in DC with a thousand other places to go;
Gonna swim across the Atlantic, 'cause that's the only place I can go!
Bad Brains, 1982

Acknowledgments

No book about DC or Parisian hardcore would be complete without a complete (as possible) acknowledgment of everyone who was a part of it along the way. Derek Alderman supervised the research and doctoral dissertation from which this book grew, and he continues to be a great friend and colleague. Additional thanks are in order to Tom Bell, Leslie Gay, and Micheline Van Riemsdijk for their notes and guidance on the original iteration. Too many other Knoxville, Long Beach, and DC friends and colleagues were indispensable in their support, but hopefully you know who you are. Of course, thank-you to my family for supporting me, believing in what I was doing, and not asking too many questions.

Thanks are in order as well for my supportive (and patient) editor at Palgrave, Joshua Pitt, a great example of what happens when a punk fan grows up and gives back. I also truly appreciate Joanna O'Neill for connecting me to the Palgrave family in Melbourne, as well as Sophie Li for her help in making this book a reality (with a cool cover).

And finally, I thank the musicians, promoters, archivists, zincsters, photographers, and everyone in between on both sides of the Atlantic who made this history. Special thanks are due to Philippe Roizès for being a great international collaborator, fact-checker, and liaison, as well as to Ian MacKaye, Nichole Procopinko, and the Dischord staff for digging through so many old letters and files. However, their contributions from Paris and DC, respectively, are just the beginning. In Paris, Lyon, Rouen,

Marseille, and elsewhere French via email, I thank: Steph Rad Party (Stéphane Delevacque), Gaël Dauvillier, Natasha Herzock, Mathieu Gélézeau, Benjamin Pothier, Roman Jaskowski, Nicolas Gresser, Nick Canon (Yannick Gaume), Nabil Ortega, Gabrielle Casseville, Florian Pons, Anne Marie (Maïe) Perraud, Olivier Firminhac, Claire Samant, Charlotte Lobert, Laurence Estanove, Philippe Beer-Gabel, Leanne Clark, Fabrice Le Roux (and the rest of Thrashington DC, in hilarious fashion), Manu Casana, Maxime Charbonnier, Norbert Mension, Arnaud Gabelli, Hugo Maimone, Philippe Cadiot, Noémie Ventura Rimmer, Momo Disagry, and Maz (Jean-François Moulard). Because the first iteration of this book originated as a study on French perspectives on Washington, most of those whom I interviewed were in Europe. However, everybody in DC and in the DC diaspora with whom I connected (and reconnected) while this book came to pass were incredibly helpful: Cynthia Connolly, Mark Andersen, Craig Wedren, Stuart Hill, Guy Picciotto, James Schneider, Ted Niceley, John Davis (University of Maryland Zine Archive), Michele Casto and Lauren Algee (DC Public Library Punk Archive), Scott McCloud, Johnny Temple, Jeff Krulik, Martin (Kim) Kane, and Alec Bourgeois. I am forever in your collective debt, and apologies to the handful of you who were left out due to editing and space constrictions. I'm grateful to know your stories and to be able to tell them here.

Roman Jaskowski (L) with the author (R), Roman's kitchen, Malakoff, France, Summer 2015

Contents

1

Introduction: A Tale of Two Cities and Scenes

On January 20, 1969, a little-known band led by Yardbirds guitarist Jimmy Page played a hastily booked show a few miles north of Washington, DC, at the Wheaton Youth Center. Or, they didn't. Accounts of this apocryphal early Led Zeppelin show are still contested. Filmmaker Jeff Krulik, known for his 1986 guerilla-style documentary *Heavy Metal Parking Lot*, investigated the alleged show in his 2012 documentary *Led Zeppelin Played Here*. He suggests that the show did happen, but several DC-area baby boomers cast reasonable doubt. After all, nobody has found any surviving flyers for the show. Nobody has any ticket stubs. Nobody, even the biggest Zeppelin fans Krulik could find, had any autographed items from that night.

All of these curiosities had reasonable explanations. Led Zeppelin's first album had been released stateside one week prior to the show and with relatively little promotion. Nobody in the audience would have known any of the songs Page's new band played. One or two people featured in Krulik's documentary claimed that the band was promoted as "the new Yardbirds." While rock 'n' roll had been a global phenomenon for almost two decades, the concert industry of established clubs, promoters (flyers),

© The Author(s) 2019
T. Sonnichsen, *Capitals of Punk*, https://doi.org/10.1007/978-981-13-5968-2_1

and distributors (tickets) was embryonic at best. The well-documented excesses of the musicians, coupled with their rigorous touring and recording schedule, did not lend much credence to their accounts. When Krulik asked Page about it at the 2012 Kennedy Center Honors, the gracious guitarist said it was possible, but he had not thought about it in decades.

Led Zeppelin Played Here may not conclusively answer the question, but it does explain to the contemporary viewer *why* one of the first shows by one of rock's most popular bands has faded from memory. While Krulik's justifications are valid, I would argue that they are more matters of time and, even more significantly, place. Look at the date. As the caravan bringing the British rock stars wound its way down to DC through blinding snow from the Midwest, Richard Nixon was swearing in as President. The local press and populace were understandably distracted from the goings-on at a suburban youth center. Americans were fighting a highly contentious war both abroad in Vietnam and at home against one another. The recent assassinations of Martin Luther King, Jr., and Robert Kennedy had sparked cycles of solemn mourning and raucous rioting, especially in a heavily scarred capital city.

In 1969, the US capital found itself in a precarious position, a button-down city licking its wounds after being torn to shreds the previous April. Across the pond, the French, whose colonial occupation of Southeast Asia had stirred into a multinational war effort, watched Parisian student protests flare into a national shutdown the previous May, the events of which theorists still analyze, celebrate, and condemn in similar breaths. It was understandable that popular culture—film, television, and music especially—was not dominating the mainstream discussion of either town. DC native Marvin Gaye, who would register one of the most beautiful screeds against the Vietnam Era in 1971, had left for Detroit over a decade earlier. Though both cities had produced a handful of noteworthy (if not commercially successful) bands over the prior decade, neither possessed anything that amounted to the status of a "rock 'n' roll town." To most of the world's population, even many native or transplanted Washingtonians and Parisians, they still don't.

*   *   *

On May 6, 2010, a four-piece from Northern Holland named Rush'n Attack played a bar basement show at Le Pix, a charming bar on Rue Pixérécourt in Paris' Belleville neighborhood. The Parisian band Kimmo, whose members booked and promoted the show, opened for them. A friend of mine from back in DC put me in touch via email with Mathieu Gélézeau, one of Kimmo's two singer-guitarists. I was scheduled to land at Beauvais airport far outside of Paris on the evening of their gig. I had never been to Paris before, and I would not have an active phone while I was in France. This would require some strategy. On April 21, I emailed the band:

> I'm a punk fan from Washington DC who wants to come check out your show at Le Pix on May 6. My buddy Ryan (who's lived in France for a little while) told me about your show, and it's happening a weekend when I'm visiting Paris, so, awesome.
>
> I'm landing at Beauvais airport at 1840h so I'll hopefully be getting to Paris around 2000h? If I get right on the metro and come out to Telegraphe should I be able to make it in time to see enough of the show?

A few days before the show, I got a reply from Mathieu:

> Hi Ty,
>   Mathieu from Kimmo here. Happy to know we have a fan in DC!
>   So, for the show at Le Pix, I think we will play at 21:30h.
>   If you are in Beauvais at 18:40h, maybe you can be at Le Pix at 21:30 but you have to be speed! If you are in Paris at 20:00 or 20:30, it will be OK.
>   We play 45 minutes I think.
>   Hope to see you at this show.

I did not reply and admit that I'd never actually heard his band's music, but his friendliness and eagerness for me to see them perform made me ready to be a fan. After catching a well-timed shuttle from Beauvais to Porte Maillot in eastern Paris, I hopped on the Metro and navigated to Télégraphe station, the closest one to the venue. At the time, I knew nobody in Paris and spoke no French. My bulging backpack gave me away as a stranger, especially on that quiet Thursday night in what appeared to be a gentrifying neighborhood (a suspicion that eventual research and better acquaintance with Paris would later verify). At the time, I was working full-time at a public relations firm in Washington, DC, enjoying one of my two luxurious weeks of paid vacation in Europe that my Parisian counterparts would likely have either laughed at or pitied. I was disenchanted with, to lift from The Clash, "working for the clampdown." I had been thinking about going to graduate school for geography, but I still did not have a comprehensive grasp on the eclectic subject, much less how prominently geography would play into what I was about to experience, tucked away on a Paris side street.

Kimmo guitarists Mathieu Gélézeau and Natasha Herzock sound check before a gig at Le Pix bar, May 6, 2010. (Photo by the author)

I arrived as the first of the three bands were finishing their set. I asked for Mathieu, and we were able to chat for a few minutes before Kimmo went over to the small stage to set up. For the few minutes leading up to their first song, I wandered over and thumbed through the items on the merchandise table, got a beer at the bar upstairs and brought it down, and stood against the back wall, trying to blend in. Listening to a cacophony of French for the first time in my life thrilled my tourist brain.

Kimmo opened their set with a song off their new album *Bolt and Biscuit*. I had never heard the song or even anything by Kimmo before, and I was thousands of miles from DC. But even before their first song ended, I felt surprisingly at home. The reason hit me surprisingly quickly: *Wow… this band seriously sounds like they're from DC.*

*   *   *

You may be wondering (understandably so) why a story about hardcore punk would begin with an anecdote about Led Zeppelin. Two key yet heavily contrasting reasons come to mind. First, the idea that any show could happen this apocryphally in DC predates the strong sense of community and the archive that would come to characterize the city's underground scene. Second, it provides a case study on the dearth of "rock 'n' roll culture" in the US capital that eventually gave rise to such a vibrant punk underground, as much as DC hardcore (or, "harDCore") did not develop in a vacuum. Bands integral to what mainstream radio would later tag as classic rock may not have actively participated in punk culture, but they formed the adolescent soundtrack to many who would build punk scenes. A pack of teenage skateboarders at Wilson High School in northwest DC listened to artists like Ted Nugent and Foghat because, until the Ramones and Sex Pistols showed up, their arena rock was the hardest, fastest, and loudest music available (see Andersen and Jenkins 2001).

Punk rock, given the cultural sea-change that it introduced, added little to the canon fundamentally. The bluntest evaluations of punk cite how the movement hit the reset button on rock music, dialing it back to its raw roots. Many characteristics of punk that become standardized after the late-1970s explosion were not revelatory. In certain respects, the mores of DIY (Do-It-Yourself) and underground, alternative music venue promotion have always existed. Punk just mollified the concept

and brought it more to the forefront of the discussion on cultural production. It sought to break down the hierarchy associated with artistic
creation as well as the socially and corporately constructed divisions
between the artist and the performer. A short list of DC artists and bands
across numerous scenes—hardcore, Go-Go, salsa/merengue—all lived
up to that ethos. DIY also sought to operate outside of the culture industry, not conforming their output to satisfy major labels' expectations.

Before rock clubs became more common in the late 1960s, rock 'n' roll
bands played wherever shows could be set up. Early rock 'n' rollers performed in high school auditoriums, grange halls, bowling alleys, and ball-
rooms. The pre-fame Beatles honed their chops playing at seedy bars in
Hamburg's Reeperbahn district (see Inglis 2012; Fremaux and Fremaux
2013), a highly valued chapter in their anthology and mythology. As Led
Zeppelin and their ilk ushered in a more proficient club industry in the
1970s, punk and the associated underground emerged as a cosmic equal-
and-opposite reaction. Ian MacKaye, one of those Wilson High School
skaters who would eventually help grow the city's punk scene to global
recognition, told the fanzine *Comet* in 2001:

> Punk has no single definition... but to me it has always meant the under-
> ground, the place where conventional approaches to life can be taken to
> task. I don't think of it as so much of a 'movement,' rather a constant paral-
> lel world that has been around for as long as there has been an 'overground.'
> There has to be a place where profit and popular tastes don't dictate cre-
> ation, otherwise we would never go forward.

MacKaye's sentiments, which he has echoed consistently for over three
decades in his roles as punk musician and head of Dischord Records,
reflect central theses on the "culture industry" which theorists like Antoni
Gramsci, Theodor Adorno, and Max Horkheimer set forth generations
ago. All three demonstrated in their heavily cited works that "the idea of
alternative culture demands an understanding of the hegemony of the
mainstream...constructed to keep the working classes and other marginalized groups in a state of perpetual disempowerment" (Spracklen 2014,
254). The existence of an "alternative" implied either outright inferiority
of the mainstream or at least holes which those on the fringes aimed to

fill. As Charles Fairchild posited, these weren't so much issues with music, but the institutions: "The creation of an alternative requires the negotiated maintenance of a specific set of social and cultural strategies while carefully avoiding the stubbornly persistent calcifying aspects of tradition" (1995, 26).

This dynamic tension, at least for the harDCore scene and its diasporic elements, has persisted for decades, despite brief flashes where the underground and mainstream have converged under the aegis of corporate interests. In the early 1990s, the commercial success of the northwestern rock band Nirvana transformed independent labels into a veritable farm system for the majors. A handful of DC-area punk musicians eventually saw financial spoils out of this transition. Shudder to Think, Jawbox, and The Dismemberment Plan all either recorded or released critically acclaimed yet commercially unsuccessful records on major labels. Jawbox publicly regretted their decision to leave Dischord, or at least to sign with Atlantic, despite being given ample creative freedom and self-sufficiency (O'Connor 2008, 24). The Dismemberment Plan was signed and then quickly dropped from Interscope Records in 1998, so they decided to release their last two albums (1999's *Emergency & I* and 2001's *Change*) on their DeSoto, a homemade label run by members of Jawbox. However, only Dave Grohl (who left the Northern Virginia band Scream to join Nirvana in 1990) and Henry Rollins (another one of the skater kids from Glover Park who left for LA in 1981 to front Black Flag) have maintained mainstream celebrity status, for which both had to leave DC. Cult figures like Shudder to Think vocalist Craig Wedren have achieved success in the entertainment industry, scoring television shows and movies without losing sight of their DIY roots. Others still, like scene kid Damian Kulash, designed occasional 7-inch record sleeves for the Arlington-based Lovitt Records before winding up in Chicago and forming viral-music video auteurs OKGO. Though some members of the harDCore family have left DC and entered the mainstream, the DC scene continues to offer an ideal of resistance and self-determination in a society that exponentially devalues art and activism.

The persistent and often tense conversation between the mainstream and underground will provide a foundation for this book moving forward, inspiring the conversation about how indispensable punk culture has always been to understanding urbanism. This book recognizes the impossibility of drawing clear-cut lines between what qualifies as "underground" and "mainstream" two decades into the realm of internet-mediated society. Though the internet was not the reason that the Kimmo show took place, it played a vital role in promotion and led me to the show, and after some years of reflection, this story.

Bibliography

Andersen, M., & Jenkins, M. (2001). *Dance of days: Two decades of punk in the Nation's capital*. New York: Akashic Books.

Fairchild, C. (1995). "Alternative"; music and the politics of cultural autonomy: The case of Fugazi and the D.C. Scene. *Popular Music and Society, 19*(1), 17–35.

Fremaux, S., & Fremaux, M. (2013). Remembering the Beatles' legacy in Hamburg's problematic tourism strategy. *Journal of Heritage Tourism, 8*(4), 303–319.

Inglis, I. (2012). *The Beatles in Hamburg*. Reverb. London: Reaktion Books.

O'Connor, A. (2008). *Punk record labels and the struggle for autonomy: The emergency of DIY*. Lanham: Lexington Books.

Spracklen, K. (2014). There is (almost) no alternative: The slow 'heat death' of music subcultures and the instrumentalization of contemporary leisure. *Annals of Leisure Research, 17*(3), 252–266.

2

DC and Paris: Capitals of Punk

This book intends to make strides in the conversation between music, geography, and contemporary urbanism through one of underground music's most storied legacies. The focus is the circulation of DC punk and hardcore music between Washington and Paris, with the lessons advancing larger discussions of sense of place and the indispensability of popular culture in understanding these phenomena. While the Washington, DC, punk and hardcore scene has been archived and mythologized expansively (see Connolly et al. 1988; Andersen and Jenkins 2001; Azerrad 2001), a negligible amount of such documentation exists of its global impact or of its valuable counterpart in Paris. Drawing histories and meanings from ethnographic interviews and textual, archival research, this story will illustrate how and why punk circulated between the two cities and continues doing so. This story will also provide a greater framework through which to understand music and the nature of urban circulation.

Like many authoritative books on any subject, the process has been iterative and extends well beyond the period of my life as an academician. The raw research directed specifically for this book took place over the past three years, highlighted by a month of in-person interviews in France with many informants in July 2015. I also incorporated two brief trips to

© The Author(s) 2019
T. Sonnichsen, *Capitals of Punk*, https://doi.org/10.1007/978-981-13-5968-2_2

DC into the project, though the project did not necessarily focus on DC's impressions of Paris, but the other way around. The following chapters will account for that juxtaposition, laying out my primary and secondary sources, acknowledging the obstacles to charting the history of this cultural circulation as well as decisions I had to make in the process. This will also include an explanation of the mechanisms of circulation (music, writing, and adapted technologies) as they both created conduits for culture to flow between the cities and have provided me with data at this stage in their life span.

This book also includes a contextual history of Franco-American cultural circulation, which predates the existence of either country under their modern constitutions. The late twentieth century is the main focus, however, as the pre-internet era in punk has become increasingly mythologized in international circles. Discrepancies exist between collective sense of place among punk fans and the realities of these places. These discrepancies reveal the nature of cultural perspective on place and collective memory, and it is important to ask why the sounds and images of the pre-internet era (roughly before the mid-1990s in the West) have dominated the discourse. The history will be told as a story of circulation, detailing both some moments in the coming-of-age and global expansion of DC punk as well as in the stories of Parisian punks who've either visited DC as musical tourists or use that city's legacy as inspiration for their own musical and artistic operations. Much focuses on these Parisian and otherwise French perspectives on this cultural exchange, as manifested through the "tourist gaze" (see Urry 1990), reflections upon revanchist urbanism (or gentrification; see Smith 1996) in both cities, and impressions of "Americanness" versus "Frenchness" in DC's alternative iconography.

Ultimately, the goal is to seek the answer to that pivotal question I asked myself in that Paris pub in 2010: What *is* it about Washington, DC? What can the Franco-American circulation of punk culture teach us about how we study alternative landscapes and understand underground tourism? How can these lessons be applied to the greater study of circulation and the greater study of popular music?

Geography is a notoriously amorphous field. Many of the overarching themes that govern scholarship on music and culture are equally applied to geography, anthropology, sociology, and musicology. However, circulation is the most relevant interdisciplinary platform through which to discuss how punk facilitated a subcultural relationship between Washington and Paris. Circulation has been the prevailing concept guiding the history of music as well as the contemporary proliferation and redevelopment of music in the internet age. The circulation of music between the Middle East and the West modified the very instruments through which the music was performed, and geographic and social particularities contributed to how musicians have modified these instruments to suit a vast field of musical needs (Wetzel 2012).

By focusing on the networks through which this music and related social practices circulated, this book seeks to elicit a greater understanding of how and why these urban landscapes affected one another. As Holly Kruse (1993) wrote, "alternative music scenes across the country and even across the Atlantic are connected rather abstractly through shared tastes… and quite concretely through social and media networks (p. 34)." Furthermore, scenes are never static, and despite the best intentions of some insiders and outsiders who cling to one proscribed imaginary, the changing positions and identities of those involved have a dramatic impact. The punk scenes in DC and Paris have both provided a platform for new participants to build upon the existing social structure. Fortunately for DC, most progenitors of the scene have aged gracefully and embraced change that newcomers have wrought.

In proposing a circulation model for musical geography, one cannot overlook the musical, geographic, and sociological realities in place at any given time in question. It would be irresponsible to deny that most circulations are uneven. Though these things cannot easily be quantified, harDCore has had a stronger impact and influence on Paris than the other way around. Additionally, circulations are not constant or even regulated. They can operate at a glacial pace (e.g. the slow burn of Minor Threat's increasing global popularity 35 years after their final show) or at an immediate pace (e.g. musicians collaborating over FTP or in real time over Skype). Circulations also stop and start at the whims of people acting within them and rarely follow a set schedule.

Another factor that contributes to spatial and social unevenness of circulation can be that which segregates them along demographic lines. Though these are not the focus of this book, race, gender, sexuality, and class must be considered when conceptualizing any circulatory model that centers on human interactions. Music scholars like Simon Frith have written extensively on how musical subcultures often collide dynamically with dimensions of class and ethnicity. As Simon Reynolds wrote for *Pitchfork* in 2018, class has always figured prominently, for example, in acceptance of musical technology and democratization inherent in that techno-culture.

Though the DC punk scene has profoundly contributed to the advancement of inclusive dialogues and discourse, one cannot ignore how Western punk culture is overwhelmingly white, young, middle-class, and heterosexual male. To be fair, however, the stereotypical image of the punk as someone who fits into all of these categories is the product of simplified mainstream depictions, a consequence of major labels' marginalization of non-male and non-white for generations. One of the ironies here is that many historic spaces of punk culture in DC and Paris have operated within and adjacent to prominently black landscapes. Geographers have addressed this phenomenon in light of other musical styles, such as Gibson and Connells' citation of blues as a geographical function of racial segregation in Memphis (2007, 178). This book will address that dynamic within the circulation of punk between DC and Paris as a historical marker for revanchist urbanism. This hegemony contradicts punk's central ethos of rebellion, which deconstructs social roles and responsibilities in the neoliberal age, thereby reinterpreting those social worlds (Thornton 1996). Cultural meanings and exchanges have different meanings to different people. Any lopsidedness or uneven pacing of this circulation between DC and Paris, however, does not detract from the fact that this cultural interaction is still very active and deserves continued attention beyond the stories told here.

It's also important to remember that punk, as a commonly understood subculture, has only existed for about four decades. It would be impossible not to acknowledge the networks already in place for centuries between DC, Paris, and their respective countries. Paul Adams (2007) refers to this feedback loop of readings and writings as "reverberation," which he sees as "a fundamental component of democracy." Of course, the relationship behind that reverberation changes nearly constantly. Much of Adams' discussion focused on the 2004 US presidential election, when anti-French sentiment was at a fever pitch in the US, at least given the rhetoric of George W. Bush and his administration's interventionist and colonialist geopolitical agenda.

Around that time, French president Jacques Chirac, a controversial and pivotal figure in French politics tracking back to the late-1970s punk explosion, was a vocal opponent of the US-led invasion of Iraq. His presence loomed over the circulation of Parisian punk culture throughout his career. In the mid-1980s, as mayor of Paris, he was often blamed for closing down the handful of clubs that nurtured the city's hardcore scene. In the mid-1990s, then president of France, Chirac, snapped at Israeli security forces when they applied excessive pressure to deflect Palestinian journalists. "This is not a method, this is a provocation," he yelled. "Do you want me to go back to my plane and go back to France? Is that what you want?" A decade later, Thrashington DC, a hardcore quintet from Brest, would lift this notorious quote and paste it over a grainy image of the White House on a T-shirt for their only American tour. Unfortunately, all they could fit on their badges was their name, free from context, which confused this American visiting Paris in 2010.

Fab Le Roux, at home in Rouen, with an old piece of Thrashington DC merch. (Photo by the author, July 2015)

To present a detailed, comprehensive history of the DC and Paris punk scenes that led up to my moment of epiphany would be impossible. Following Kevin Dunn's example in *Global Punk* (2016b), this book, out of necessity, eschews any claims to comprehensiveness. This is by no means a definitive history of the DC punk scene or its counterpart in Paris. More comprehensive variations on both, however, are available and recommended. Mark Andersen and Mark Jenkins collaborated on *Dance of Days* (2001, Akashic Books), the cornerstone reading on the twentieth-century history of punk in the nation's capital. Photographer Cynthia Connolly released the quintessential photo-document of American hardcore: *Banned in DC* (Connolly et al. 1988), now in its seventh printing. Many writers have profiled Minor Threat in popular-press books, including Michael Azerrad in *Our Band Could Be Your Life* (2001) and Andrew Earles in his collection *Gimme Indie Rock* (2014), the latter of which featured a photo from a Minor Threat show on the cover. Long considered hardcore standard-bearers, images and iconography associated with Minor Threat and their predecessors Bad Brains have permeated literature, artwork, and record labels. One French label we will be meeting later, Crapoulet, uses a reconstituted photo of a young, screaming Ian MacKaye as the header image for their Bandcamp page.

Though this book focuses primarily on underground scenes in Washington, DC, and Paris, the amorphous concept of the "mainstream" must be understood to conceptualize the "underground." One cannot exist without the other. The omnipresence of popular culture in urban settings creates a forum for resistance for many. The underground cannot exist without whatever figures, icons, and forces dominate above the surface, just as the "mainstream" inherently necessitates an opposite (if not quantifiably equal) reaction.

For those who actively identify as part of an underground scene, much of what defines their place therein and accrues subcultural capital is an active refutation of sweeping assumptions about the mainstream. Sarah Thornton (1996) wrote about this dichotomy within the British Club scene of the late 1980s and 1990s, implying that a lot of the underground grows around what it isn't, rather than what it is. Where involvement in the scene is rarely, if ever, financially lucrative, symbolic and social capital (see Bourdieu 1984) become the thickest currency. For those who aligned themselves socially and politically with harDCore, both in its salad days or more recently, these dichotomies are obvious. The mainstream was corporately controlled and globally oriented; harDCore was independent and locally oriented. The mainstream was mediocre, passive, and uncritical; harDCore was innovative, aggressive, and highly politicized.

While this book asserts that harDCore has changed the world, albeit at a glacial pace, the grandiosity of the sentiment still belies the reality that none of these bands are household names. Although Bad Brains has been nominated to the Rock 'n' Roll Hall of Fame and Minor Threat T-shirts are available at many Urban Outfitters stores, their fans are still in the minority of global citizens. Appropriately, few denizens of these scenes harbor any delusions of grandeur; some continue making music at their own pace and by their own rules, and others have led mostly nonmusical lives.

Anyone unfamiliar with American hardcore punk (and the French circulation and reinterpretation thereof) would be wise to look to earlier stories and oral histories for anecdotal and visual accounts of the culture. It would be wiser yet to sample the music of these cities and of this era to hear the soundtrack of the underground themselves. Curious listeners today, unlike the people who made the music at the time, are in the fortunate position to find almost any punk song of renown on YouTube, Spotify, or similar digital platforms. No longer does one need to know somebody, no longer does one need to wait months for an unreliable distributor or label to mail an import 7″ (though delays like this are still possible with some labels), and no longer does one need to rely upon carefully crafted mix tapes and mix CDs to discover new old music. Dischord, Lovitt, Prohibited, and other localized "folk labels" from DC and Paris have made their catalogs available for streaming and purchase via the website Bandcamp. Some have been diligent about keeping their back catalogs available on vinyl, too. Early pressings of Minor Threat records that draw four-figures on eBay and Discogs are more the exception than the rule.

Because this project celebrates and details the exchanges of underground culture between DC and Paris, a wide array of bands from both cities appear within the narrative. Of course, considering how heavily the research relies upon extant documentation, the best-known bands from each city may dominate the spotlight. Particularly when discussing the mainstreaming (culturally and politically) of indigenous punk scenes in the respective towns in the late 1980s and early 1990s, bands like Fugazi and Bérurier Noir usually occupy the foreground. Especially in the case of Fugazi, the band is likely the most documented acts in the DC punk canon, formulating what rocker Ted Leo once referred to as "the paradigm of the successful indie rock band." Therefore, they will figure predictably and prominently in subjective tracings of the Franco-American circulation of DIY punk culture. But, as the bootlegged T-shirt says "This is not a Fugazi t-shirt," and this is not a Fugazi book.

None of this should minimize the accomplishments and contributions of many other bands from DC's punk era, certainly not limited to the Fugazi members' previous bands like Minor Threat, Rites of Spring, Embrace, One Last Wish, or Happy Go Licky. Too many other seminal bands within the scope of new wave (The Slickee Boys, Urban Verbs, et al.), hardcore (Bad Brains, Scream, Void, SOA, et al.), post-hardcore (Jawbox, Shudder to Think, DC3, Fire Party, et al.), and later, indie pop (Q and Not U, The Dismemberment Plan, et al.), pop-punk (Fairweather, The Max Levine Ensemble, et al.) and ska (The Pietasters, Kill Lincoln, et al.) operated within and outside the Dischord Records orbit to attempt anything close to a comprehensive scene history here.

In fact, Bad Brains, often considered the trailblazers of the fast and loud DC hardcore sonic aesthetic, did so in the years preceding Dischord's founding and never released anything on the label. Bad Brains moved to New York early, but are permanently considered a DC band. Bad Brains' legacy is a cogent reminder that though the Dischord stable and scene enveloped most of the music that influenced international impressions of DC, the label's catalog is hardly a comprehensive telling of the story. Several other labels, including The Slickee Boys' Dacoit, Skip Groff's Limp, Jawbox's DeSoto, Unrest's Teenbeat, and turn-of-the-century Arlington mainstay Lovitt, served the regional music scene in their respective ways over the past 40 years.

The Paris punk scene, however, does not have as recognized of a legacy. Many American bands, particularly ones from DC, have been immortalized in coffee table books and rockumentaries, including Jem Cohen's *Instrument* (1998) and more recent entries like Scott Crawford's *Salad Days* (2015) and James Schneider's forthcoming *Punk the Capital*. This level of documentation has not been the case for most of mainland Europe. Little academic attention has been afforded Paris as a punk rock city, and next to none has been devoted to the city's hardcore punk underground. This is one of the great inequities that French punks have, unfortunately, grown to accept over the years. Cultural historians have foregrounded this lacking in studies about everything from French underground newspapers, zines, and fashion, and I'll reiterate it here. Paris had most all of the hallmarks of a proper cradle of punk culture—the teeming Les Halles neighborhood, clubs like Gibus and shops like Harry Cover, the Open Market, and New Rose (see Warne 2013)—but no French bands enjoyed anything close to the success of their London and New York counterparts in the late 1970s. Outside of some provincial bands who crossed over into new wave success, most faded into obscurity.

Outside of Christian Eudeline's often-maligned 2002 compendium *Nos Année Punk*, few formal texts have emerged anthologizing the French punk movement. Some books on the history of French rock (e.g. Médioni 2007) include bands from the 1970s and 1980s affiliated with the punk movement, but only a handful of commercially successful ones like The Dogs (who figure into the DC punk story) and anti-Fascist agitators Bérurier Noir. The underground is easily lost in the balance.

Washington, DC, as Punk City

What's wrong with DC? Why is it so easy to ignore? The town you love to hate? The obsession that died: Too many timid bureaucrats? Transient government employees? All political power and no pretty power? Not enough clubs? No money-sucking labels? No Ego? No college radio uniting the area? CIA plot to clean sweep the streets of Main Street, USA, of subversive influence? Lack of diverse financial base? Bad Fed reputation? Too many idiots comparing it to bigger media meccas? (*Mole* Zine, No. 6, 1994)

Remember! The scene you crave should be the one you create!!!!!! (*Capitol Crisis* Issue #1, 1980)

While it may be simple to disregard DC as an aberrant urban system in the shadow of American megacities like New York, Chicago, and Los Angeles, this chapter takes the opposite stance. Washington, DC, while its federally oriented origin may set it apart from traditionally developed urban centers, has undergone generations of growth, development, inequality, and loss that reflect the greater anxieties of American culture. The federal government has not lost any significance, but DC's growing service-sector economy "might be very telling of future urban growth throughout the country" (Hyra and Prince 2016, xlvi). Overpopulated as it may be, the District and its surrounding Maryland and Virginia suburbs (or, colloquially, the DMV) encapsulate the warring social dynamics of any American city in a globally comprehensive context. Helen Taylor (2001) wrote that cultural diffusion must be understood transatlantically, and I argue that the urban landscapes that provide the spaces and places for these documents to become monuments (as Foucault put it in *The Archaeology of Knowledge*) must also be understood in a global framework.

Landscapes like those of DC and Paris, often "constructed with specific ideals in mind… by ideologically driven governing elites" (Hoelscher 2009, 137), have come into greater focus in cultural geography over the past decade. Usually, these analyses offer a critical deconstruction of such hegemonic narratives of cityscapes and ways to democratize place and space. As Hoelscher (2009) continues, "even the most ordinary, everyday, and taken-for-granted landscapes carry symbolic meanings that can be interpreted for their iconographic intent and effect" (139). When I presented an early iteration of this book at the annual meeting of the Society for Ethnomusicology Southeastern and Caribbean division (SEMSEC) in 2016, my greatest epiphany came from watching the presentation that followed mine. Victor Hernandez-Sang (2016) from the University of

Maryland spoke about the international diversity of Latin-American bands in the Washington, DC, area. It struck me not only because it was a well-researched and thoughtful project, but because I remembered these salsa, merengue, and Cumbia groups within my personal landscape of DC. While punk was what made me interested in DC, I soon found a vibrant Latin-American community embedded in that city's nightlife. Even as a white twenty-something out late in the Adams-Morgan neighborhood, I vividly remember hearing three or four different bands of Latin-American musicians churning dance rhythms and call-and-response melodies out the doors of the intimate nightclubs like Bossa or Chief Ike's Mambo Room (RIP) and into the ears of passersby. These Latino bars and clubs sat mere blocks from the house where Minor Threat played their first show in 1980. The DC that Hernandez-Sang presented through the lens of immigrant salsa musicians was just as strong of a counternarrative to "mainstream DC," every bit as intriguing, and every bit as valid. DC's Latin-American music scene, while not full of tourism signposts (not yet, at least), actually exemplifies the "e pluribus unum" ("out of many, one") melting-pot spirit of America, while overcrowded monuments downtown merely regurgitate that phrase on nationalist symbols. DC's Latin underground, similar to the punk underground, has accumulated little mainstream fanfare, though its quotidian role in the urban landscape confronts the hegemonic narrative that draws millions to the National Mall each year. As geographer and semiologist Owen Dwyer might term it, one cultural landscape pales in symbolic accretion to the other, more established one, though both possess it in no uncertain terms.

On the mainstream level, DC's most prominent structural icons owe an obvious aesthetic debt to European architecture, with an emphasis on French ideals. The White House is likely the most famous building in DC, serving as geographic shorthand for both the federal city and the US in film, television, and elsewhere within the pale of popular culture. As later chapters of this book will address further, international audiences notice this simplification.

Genres like punk and hip-hop, both of which notoriously emphasize and prize the "local," function as unwitting geographic mile markers in the timeline of their urban landscape. Kevin Dunn tackled this extensively in *Global Punk*, citing scholars such as Martin Stokes (1994) and Mark Olsen (1998) who both expounded ideas on the geographic value of music, the latter characterizing scenes as "'territorializing machines' that produce particular kind of relationships to geographic space" (Olsen 1998, quoted in Dunn 2016b, 65).

Ian MacKaye uses the term "regional accent" to describe the collectively read essence of a city and connected scene. A cursory run through a set of noteworthy releases by DC bands has lyrics and artwork that directly represent their town. In 1991, Nation of Ulysses sang songs like "You're My Miss Washington, D.C." and "Hot Chocolate City" in frenetic tributes. Fugazi's third full-length *In on the Kill Taker* (1993) featured a hazy photo of the Washington Monument on the cover. Q and Not U titled an early single "And the Washington Monument Blinks Goodnight" (2000). The Dismemberment Plan's song "Spider in the Snow" features singer Travis Morrison memorializing his young-professional experience of "[walking] down K Street to some temping job" (Morrison 1999). Gray Matter's song "Oscar's Eye" and the title of its host album *Food for Thought* (1985) were both named after multi-purpose venues within DC that had been vital to the scene's development. Their contemporaries Rites of Spring titled a song "Hain's Point" (1985) in honor of the scenic southern outpost on Roosevelt Island. The Capital City Dusters named their second album *Rock Creek*, after the vast green space that slices through the center of the District.

Suburban bands made similar marks, expanding the geographic signature of DC punk to well outside District lines. Early pressings of Scream's 1982 *Still Screaming* LP (the first full-length album that Dischord released) have a prominent 'bXr' on the labels, a nod to Bailey's Crossroads, the Northern Virginia community where the band lived. Hoover's 1993 album *The Lurid Traversal of Route 7* refers to the busy state road that led into and out of DC from Winchester, Virginia. Even the noisy, experimental quintet Black Eyes had a song called "King's Dominion," paying tribute to a popular amusement park down Route 95 in Virginia that has been a longtime day trip destination for DC teens and young adults. The list goes on.

Paris as Punk City

> Few things in the history of humanity are as well known to us as the history of Paris. Tens of thousands of volumes are dedicated solely to the investigation of this tiny spot on the earth's surface. (Walter Benjamin, *The Arcades Project*, 1940)

The role of Paris in any discussion about global culture and Western society needs little justification. Walter Benjamin famously named it "the capital of the 19th century," where high and low culture converged and modernism and capitalism both became refined as art forms unto themselves. Few countries in the Western world are as emblematic of "culture" in the intellectual sense as France, given the country's unparalleled roster of nineteenth- and twentieth-century philosophers. France's political epochs have defined much of the global political imaginary, including the by-products of the French revolution that gave us "the left" and "the right." Any attempt to distill French or even Parisian culture into a set of pages would be impossible but would also belie the point about punk's role within that greater scheme. Music scholars have accordingly situated Paris as the rare urban paradigm of these ideas.

As early as 1855, the poet Charles Baudelaire was noted for using the term "Americanized" in a negative light regarding the creeping US influence on the World's Fair (Green 2014). Credited with coining the term *modernité* (modernity), Baudelaire fetishized Paris' sordid landscape, arguably laying the groundwork for art which celebrated and canonized the city and urban ethos 120 years before Metal Urbain brought Paris into the global punk conversation with "Panik." Baudelaire's archetypal character of the flâneur, an artist and passive observer of the overstimulating city, influenced generations of thinkers. Benjamin, though not native to France, centered most of his most influential work around Paris. These included *Das Passangen-Werk* (The Arcades Project), a collection of writings about the prior century of creeping urban phenomena that was unfinished at the time of his death in 1940. The recently deceased Polish sociologist Zygmunt Bauman contrasted their approach, hinting at several tropes which would inform the late-1970s punk explosion (e.g. "the modern world" for The Jam and the Buzzcocks' "boredom"):

It is the modern world which is the original flâneur, the Baudelaire/ Benjamin human flâneur is but its mirror image, its imitation, the product of stock-taking, of forced adjustment and mimicry. Like the world which is his home, the flâneur wanders without aim, his stroll punctuated every once in a while by looking around. Without aim? That aimless stroll is the aim; there could not be, there should be other aims. (Bauman 1994, 139)

In the early 1950s, French philosopher Guy Debord became a proponent of the dérive, where one operated at the whims of the landscape he or she encountered on his or her stroll. In another case study of Franco-American refraction, American city planner Kevin Lynch, almost simultaneously, was promoting the same idea of treating the city as an active agent in its own creation and modification (Long 2014, 50). Both Debord and Lynch were integral in the establishment of psycho-geography, a governing concept in the relationship of mental perception and impressions of an urban landscape both nearby and distant. As Long (2014) goes on to contextualize psycho-geography, he indicates the critical role of music and art in directing those exploring urban environments away from proscribed tourist landscapes and into alternative spaces and personal vantage points. More recently, these underground stories, rendered more accessible via the internet, have become celebrated hallmarks of against-the-grain tourist practices in both Paris and DC.

Though punk did not originally self-actualize as a rock subgenre in France, some argue that it could not have grown or proliferated without the influence of radical French thinkers. Citing Paul Yonnet, Jonathyne Briggs (2015) wrote that "punk provided a method for directly critiquing French society and engaging young people politically … [thus continuing] a tradition of critical culture that had deep roots within French history." Though he was too old to catch the wave of punk in the late 1970s, prominent twentieth-century philosopher and semiotician Roland Barthes is frequently cited in literature on punk culture. In *Subcultures* (1979), the first major academic release on punk and reggae culture in the UK, Dick Hebdige implied that punk was a natural extension of Barthes' ideals. To him, Barthes' writing radically blurred interpretations of high and low culture in late-twentieth-century mass media. To Barthes, these musical rituals and forms were often subsumed into a national folk-lore by those who controlled the media:

> The whole of France is steeped in this anonymous ideology: our press, our films, our theatre, our pulp literature, our rituals, our Justice, our diplomacy, our conversations, our remarks about the weather, a murder trial, a touching wedding, the cooking we dream of, the garments we wear, everything in everyday life is dependent on the representation which the bourgeoisie has and makes us have of the relations between men and the world. (Barthes 1972, quoted in Hebdige 1979)

Like Dead Kennedys singer and progressive activist Jello Biafra would implore his spoken-word listeners to "become the media," Barthes saw the beauty in the everyday and, as the writing on the wall often said during the 1968 French proletarian uprising, that beauty was in the street.

As the urban microcosm of French popular culture, Paris occupies a rare position in myriad media-based worlds. I would argue that les soixante-huitards (those in "the spirit of '68 unrest") presaged the advent of punk in France through their systematic rejection of the old ways and full-scale rebellion. They also worked to disintegrate barriers between the government and the people. Much like the urban riots of 1968 in the US, the student and worker protests throughout France brought much more serious attention to issues of urban unrest and poor living conditions. Much more so than the prevalent hippie culture in the US at the time, the Paris student demonstrators formed an unwitting coalition with a proto-punk art movement and the far-left thinkers of that era, many of whom are still quoted ad infinitum. Henri Lefebvre, the Marxist philosopher whose work has been widely expounded by geographer David Harvey, first published on "the right to the city" in 1968, which was no coincidence.

The legacy of May 1968 found a thinkable counterpart in the punk movement in Paris, and those tropes have reappeared throughout popular culture since then. The French crime novel (Serie/Roman noir) is a key example. Because France's punk scene is generally smaller than its Anglo-Saxon counterparts, "the roman noir… is on numerous levels punk rock for France…commonalities between punk and crime fiction are important because they break down the conceptual divisions between the social roles of music and literature" (Lee 2005, 186).

Paris' emerging and strengthening diversity during the punk era played a valuable role in the interplay between underground music and urban landscape. Before punk culture became mass-marketed and refracted through simplified representations on television and film, "punk in France acted as a powerful nexus not only for fluid and creative cultural contact, but also for political reflection" (Warne 2013, 220). An ethnic and racial diversity of characters, including the first skinheads in Île-de-France, came from all over the city and les banlieues (suburbs) mixed in at clubs like Les Bain Douches and Le Palace, changing dialogues about French identity in the process.

Punk and post-punk have inspired both music journalists and academics to weave geographic narratives into the aesthetic narratives of the music. Many of these are done posthumously and indirectly, imposing psychogeographic meaning and structure on words and music. Manchester post-punk group Joy Division, despite having a relatively small recorded output (two full-length LPs and a few singles), has been among the most prototypical examples. Even prior to the 1980 suicide of the band's enigmatic singer Ian Curtis, British journalists like Paul Morley had been ardently associating the band's musical aesthetic with Manchester's urban environment. In the past two decades since Lily Kong's outcall (1995) for greater consideration of popular music in geography, scholars in related fields have been drawing on Joy Division and other Manchester groups for inspiration. Manchester cannot exist outside of "a centuries-long development both in fiction and history books of working class and industrial connotations" (Bottà 2009, 351). As Manchester was ground zero for the Industrial Revolution, bands like Joy Division, The Fall, Buzzcocks, The Smiths and Happy Mondays have been unable to avert dirty, grim, working-class associations. Sociologist Leonard Nevarez (2013), who spent his teenage years going to underground shows in the DC area, approaches Joy Division's legacy from this critical perspective:

[The band's reluctance to speak to Manchester's impact in interviews] highlights the mediation of history, geography, social relations, and technology involved in the act of listening. It further raises the question, what are the contemporary contexts in which people might perceive Joy Division to sound like the Manchester of old? (p. 58)

The influence of Joy Division, Cabaret Voltaire, and other gritty UK urban-decay post-punk was immense on what quickly became the French Cold Wave sound of the early 1980s, "a musical symbiosis of the UK post-punk guitars and German industrial electronics that rhetorically mimicked – through dark melancholic sonorities – the acoustic contours of Europe's angst-ridden cultural landscape" (Hall 2016, 152). Today, Paris' most prominent punk bands like Frustration openly bridge this tradition with more modern interpretations. Though DC and Paris have hegemonic representations and perceptions of their urban landscape that differ from Manchester or similarly (historically) industrial cities in their respective countries, artists from both have had to countenance curiosities about psychogeographic life during their most active periods. Where the city is the canvas for creative expression in rock music, the city's changing meaning in the public memory is also crucial in music criticism. Ultimately, the changing meaning informs the growing *counternarrative* of that city, a cogent reminder of music's role in that development.

Though often finding itself perpetually "behind the curve" (as several of my informants attested in interviews), Paris has made great strides and contributions to the discourse on punk circulation. Over the years, Parisian bands like Prohibition have drawn from the jazz and art-rock inflections that motored DC post-hardcore and created sonic reflections of their city. In 1996, on the heels of a period when they shared stages with Fugazi all over Europe, Prohibition released the LP *Towncrier*, which they called a "concept album about urban life and Paris in particular … carried out as an experiment by the band." Clearly, that DC-style predilection for pushing boundaries and hometown reflexivity had found a comparable laboratory on the Seine.

Why DC and Paris?

Washington, DC, and Paris, France, are politically, historically, and culturally inextricable from one another. Both cities are prominent national capitals, rife with nationalist symbols that all occupy permanent space in the global imaginary. Every simplified skyline of DC includes the White House, the Washington Monument, the Capitol Dome, and a steady selection of others in some form (the Smithsonian Museums and memorials to Jefferson, Lincoln, Martin Luther King, Jr., are all popular). Every simplified skyline of Paris includes the Eiffel Tower, the Arc de Triumphed, Notre Dame Cathedral, and a steady selection of other, more esoteric choices (Le Louvre Pyramid and Montmartre are also common). Both cities are microcosmic of the democratic process under which so much of the world is governed. Both cities, though navigable by motor beltways, are very difficult or plain idiotic at times to take by automobile. Both cities have expansive and often-troubled public transit systems they call Metro. Both cities have been racially segregated on varying timelines and to varying, yet significant, degrees. Also, to those not mincing histories, both were designed by Frenchmen.

If DC doesn't need to rely on its musical heritage for tourism, then certainly neither does Paris. If there's one thing that neither city lacks, it is tourists. Paris received 22.4 million visitors in 2014 just to its hotels (Paris Office du Tourism et des Congress). This does not include Airbnb, Couchsurfing, and personal visits to family and friends. These visitors come from all over the world and engage in almost every type of known tourism as well, most of which overlap. Notre Dame Cathedral, for example, satisfies touristic ambitions for both religious pilgrimage and world heritage sites. Some general attractions are much more heavily sought after in Paris than in DC, but everybody has to eat no matter where they're staying. Regardless, both cities are stark examples of nuclei in the world's most developed countries that have attracted an increasing amount of attention from social scientists. As post-industrial societies lean their economies increasingly on the service sector, tourism becomes that much more of an economic anchor.

An obvious contrast between the two cities' histories is that DC was devised specifically to be the home of the American federal government, and Paris has a history going back almost two millennia. The plan for the District of Columbia was laid out by a Paris native named Pierre L'Enfant, today immortalized with a statue in his likeness and a major civic plaza (and Metro station) in his name in the city's southwest quarter. L'Enfant came to the US to assist the Continental Army under the command of the Marquis de Lafayette, also immortalized with a statue and an often-crowded park north of the White House. George Washington entrusted

L'Enfant with the duty of laying out a general plan for this city on the Potomac River, and the planner's recognition of French nobility and monarchic symbolism did not defeat the Enlightenment ideals under which his adopted country had been founded. However, the two symbolic systems did clash. Cultural Geographer Denis Cosgrove (1989) put it best:

> [Pierre] L'Enfant composed the plan… of two simple geometrical designs: the orthogonal radiating pattern traditionally favoured by European monarchs exercising an absolute power which radiated from their persons and their courts, and the infinitely repeatable grid pattern which had become the basis for every colonial town, a democratic and egalitarian form that gives no single location a privileged status. (p. 129)

The star-and-grid pattern did not anticipate the advent of the automobile, of which DC had the greatest per capita usage of in the nation during the 1920s. Nor did the original plan anticipate how extensively the city's population would expand beyond the lawmakers, their families, and those who worked for them. This cartographic signature, which followed the classic European model that no structure could be "closer to God" than the central cathedral, dictated that no building (save for the Washington Monument) would rise higher than what would soon be the Capitol spire. The Rosslyn business district across the Potomac River could circumvent these statutes by being in Arlington, but the District proper has stayed uniquely flat for a major city. While buildings on the Virginia and Maryland fringes are taller than the Capitol, few are sky scrapers like those of New York or Chicago, none of which existed during L'Enfant's time. By the end of the nineteenth century, DC had grown so haphazardly that it no longer seemed like a planned city at all. The federal government had done little to control it, and it was not until they constituted the McMillan Plan in 1901 and the Park Commission Plan in 1902 that they "modified, enlarged, and reestablished" L'Enfant's plan of Washington (Jacobsen 1965, 13).

Unlike DC, Paris is a primate city. Though France's smaller cities like Lyon, Marseille, and Nantes are all significant in this story, Paris remains the dominant core of the country's cultural industries. Paris is the epicenter of the French motion picture industry, the standard-setting French fashion industry, and the corporate presence of the French music industry. Though DC has all of those industries in smaller doses, they cannot compete with the suppliers of New York, Los Angeles, and, especially in terms of music tourism, New Orleans (see Gotham 2005) or Austin (see Porcello 2005). However, even more musical cities lacked in resources for musicians bent on making careers out of their craft, necessitating moves to cities like New York or Los Angeles. Washington's two most commercially successful punk musicians, Henry Rollins and Dave Grohl, both had to leave the DMV to attain mainstream success, the former (eventually) as a major label recording artist and film actor and the latter as drummer for the meteoric Seattle band Nirvana and later front man for Foo Fighters. Rollins and Grohl, however, are two prominent figures in a long list, including Duke Ellington, Al Jolson, and Marvin Gaye, whose early lives were spent in DC but entire professional careers were spent in, or based in, major cultural centers.

Punk and Musical Geography

> It is perhaps appropriate that the punks, who have made such large claims for illiteracy, who have pushed profanity to such startling extremes, should be used to test some of the methods for 'reading' signs evolved in the centuries-old debate on the sanctity of culture. (Hebdige 1979, 19)

Dick Hebdige, among the first social scientists to fully embrace the punk counterculture, also testifies here for how integral punk is in understanding the power of urban counternarratives. One cannot discuss circulation or musical formations without discussing culture. None of which could be addressed without the frame of geography, because "as Johansson and Bell (2009, 9) succinctly put it, 'place is omnipresent in music and, reciprocally, music is clearly evident in place'" (quoted in Long 2014).

The ethnomusicologist Bruno Nettl once wrote that "music… grew out of materials already present: animal cries, speech, rhythmic activity" (2005, 261). As a musical tradition, punk rock grew out of materials already present: power chords, defiant or didactic songwriting, feedback, and distortion. In fact, very few of the musical conventions associated with punk anywhere in the world simply appeared or just occurred spontaneously. Punk synthesized a vast constellation of existent ideas into a palpable movement. Like all genres of popular music, punk "existed before it had a name" (Weinstein 2011, 36).

Before it had the punchy four-letter word, it had spread well beyond the confining, nonrepresentational structure of music. Denizens of the confounded yet intrigued news media notably associated punk with rebellious fashion, anti-establishment attitude, and delinquency. Music only made sense as a launching point for punk because not only did "punk rock" roll off the tongue, but as musicologists have argued since well before the 1970s, music was inextricable from the quotidian life of its adherents. "There is probably no other human cultural activity," wrote anthropologist Alan Merriam in 1964, "which is so all-pervasive and which [reaches] into, shapes, and often controls so much of human behavior" (p. 218).

Merriam wrote *The Anthropology of Music* long before music was widely accepted as a theoretical launching point in the academy, but his work has remained influential throughout the humanities over five decades later. Recent research on the role of music in social development and the greater pale of human activity reinforces and builds upon these ethnomusicological theories, implying that music should be taken as seriously as, if not more than, written and spoken language. It's not a tough argument; many contend that music predates them both.

As punk music grew into commodities and marketability, it wrote a new chapter in the history of popular music. By all accounts, rock 'n' roll had largely stagnated by the mid-1970s and many popular rock writers lamented that the original narrative of rock 'n' roll had been lost. Critics, theorists, and fans all over the world all have their own opinions why: substance abuse had taken too much of a toll (e.g. Jimi Hendrix, Janis Joplin, Jim Morrison, Brian Jones, et al.), the garage rock underground which the British invasion had spawned had burned out, the US retreated from Vietnam with its tail between its legs, and the hippies had lost their idealism, among other reasons.

After punk blew onto the charts concurrent with the disco craze in 1977, writers retroactively sought alternative sources outside of the dueling ground zeros of New York and London. Detroit bands like the MC5 and The Stooges are commonly considered godfathers of fast and loud music, but much like all musical genealogy, the borders are not so easily cut and dried. Though independent labels sprang up to release the less mass-marketable sectors of punk music at the end of the 1970s, the parallels between this era and the dawn of rock 'n' roll were certainly not overlooked. Indie labels, mostly bred out of studios like Sun, are widely credited for the genesis of rock 'n' roll. Even Sun head Sam Phillips was working off a model; small labels had been releasing jazz, country, and blues to niche audiences for as long as records were an accessible format.

None of this, however, had dramatically altered bands' commercial prospects by the end of the 1970s. Even though Paris was a major city by most metrics, homegrown bands, especially those imitating punk and forging French takes on the style, had little financial foothold. So it would make sense that Washington, DC, was making even less of a blip on the radar. Young punks in both cities, however, would profoundly change that over the following decade.

Leading up to punk's commercial heyday in the mid-1970s, there was still little uniformity between the biggest markets. In New York, punk was more of a lark, taking the piss out of then-prominent excesses in arena rock culture. New York bands who appeared ahead of the curve, like the New York Dolls and The Dictators, injected 1950s-style urban(e) rock 'n' roll with an affect that fell on a spectrum between sheer glam/androgyny or (particularly for The Dictators) self-effacing humor with home-style Jewish and Italian cultural subtexts. While the Dolls, like The Velvet Underground in the prior decade, enjoyed some insular influence among New York's tastemakers like Andy Warhol, the joke didn't exactly play outside the five boroughs (or even throughout all five boroughs). The Dictator's 1975 debut *Go Girl Crazy!* remains one of rock's funniest inside jokes, referring to singer Richard Blum (aka Handsome Dick Manitoba) as "the handsomest man in rock 'n' roll," covering Sonny & Cher, and singing "we're all members of the master race / we've got no style and we've got no grace" (Shernoff 1975). Sire Records released The

Ramones' debut album the following year. While The Dictators' appeal was still present, The Ramones were even more stripped-down and appealing to curious music fans in both the Americas and across the pond who had never heard anything like Johnny Ramone's buzz-saw guitar before. Other New York bands that circulated the Bowery/CBGB's scene like Blondie, Television, and The Dead Boys (though they originated in Ohio) did not necessarily mimic The Ramones' style but were clumped together nonetheless by an increasingly eager music press. In this span of less than 18 months, punk bands popped up all over North America, tied together through an informal network of underground press, tape trading, and word of mouth. To give the hundreds of noteworthy bands a mention here would be impossible.

Meanwhile, in London, a generation of British entrepreneurs packaged punk as a stylistic statement, and a generation of British youth who felt they had no future bought it hook, line, and sinker. One of punk's great successes in the UK was putting a public face on working-class squalor, breaking down the rigid class system through music that challenged a stiff status quo. Where punk was still regarded as significant (though some derided it as a fad) back in the US, punk in the UK was more aggressively threatening the tenuous social order of England. Malcolm McLaren's project The Sex Pistols embraced all the clichés of punk, toured regionally, and purposefully confronted the establishment for a couple of red-hot years before imploding in early 1978. UK contemporaries like The Clash, The Jam, The Damned, Buzzcocks, Siouxsie and the Banshees, and an impossibly long list of others both contributed to and capitalized upon the zeitgeist. Despite the brief life span of this confluence of bands in the mainstream, this epoch remains the dominant paradigm of how "punk" is imagined and marketed by those unfamiliar with the underground.

In Paris, a homegrown scene sprouted that drew heavily from nearby London, even to a point where French punks revered British identity and the use of English as indicators of "authentic punk" (Briggs 2015, 147). This reverence, at least early on, stripped nascent Parisian punk bands of their outward Frenchness. Stinky Toys, allegedly the first self-labeled punk band in Paris in 1976, suffered from lukewarm reviews and a lack of label support. They did perform in London early in that city's punk explosion, but faded from prominence as the scene grew in both cities.

The free movement of individual agents among Paris, London, and New York throughout the 1970s makes tracing the movement of proto-punk and punk music, style, and ethics nearly impossible. Jonathyne Briggs attempted to diagram this early circulation in *Sounds French* (2015). He cites the Parisian Elodie Lauten as an ostensible "fan zero" for French proto-punk; she spent much of the early 1970s living among New York's tastemakers, bringing some artifacts of that scene back to Paris with her in 1974. According to Briggs, Lauten's "ease in moving between the punk worlds of New York and Paris reveals the cosmopolitan character of punk and certainly influenced its development in France" (p. 156). Similarly, Mark Zermati opening the Open Market shop in Paris in 1972 seems, in retrospect, like a cornerstone in the foundation of French proto-punk culture. Setting the tone for punk-centric shops like New Rose that would emerge later in that decade, Zermati imported underground American and British music, making much of it accessible for the first time on French soil. A founding member of Lou Reed's Parisian fan club "Les Punks," Zermati would also found Europe's first proper punk festival, bringing the Damned to the Mont de Marsane bull ring on August 21, 1976.

Clearly, that "cosmopolitan character of punk" contributed to the cultural development of these scenes and added a valuable chapter into their respective histories. A more detailed history of the cultural circulation between Paris and these cities will be contextualized within a history of Franco-American circulation in the next chapter.

Punk and Social Theory

The academy was generally slow to embrace the double helix of punk music and culture, but many theoreticians of the era recognized the multiplicity of intellectual movements that informed punk rock. Many rhetoricians had written extensively about protest music and associated movements (Cathcart 1972; Knupp 1981). The landmark postwar counterclutures of the 1950s and 1960s had also widely inspired those in the academy, including the beat generation (Freud 1959; Masserman 1967), hippie counterculture (Becker 1967; Hall 1968) and Caribbean-influenced rudeboys (Simpson 1955; Nettleford 1970), all of whom were bound to particular musical scenes and styles.

The first social scientist to address punk formally and enthusiastically was Dick Hebdige, an English sociologist who incorporated UK youth movements in his 1979 volume *Subculture*. In his 20s at the time, Hebdige made a critical argument that punk was a confluence of multiple counterculture mores, which were "subject to historical change, each [instance representing] a 'solution' to a specific set of circumstances, to particular problems and contradictions" (1979, 81). This fluid dynamic of punk which has appropriately aligned it with the palimpsest landscapes of its host cities will be addressed in greater detail here.

While Hebdige's "claims of illiteracy" in punk were essentialist and stereotypical assessments, one can understand where he was coming from. Thirty years after publishing *Subcultures*, he published a retrospective paper that explained the time-place context for his writings. He does not necessarily backpedal on his initial points about punk, reggae, or mod culture, but he does take time to explain his intent and the contextual realities of the project. He signed his contract in 1976 and submitted the draft for publication in 1978, which was before the post-punk era (encapsulated by funkier bands like Gang of Four and The Slits, both of whom formed in 1976 but released their first albums in 1979) cohered and blew initially dismissive critics and academics out of the water. Similar to many punks' early records and on-the-record statements (lyrics, liner notes, zine interviews, etc.), he also felt that his early points had gotten away from him and recontextualized, for better or worse. Ultimately, he felt guilty for "placing a gaudy, cartoonish wrapper on a serious activist and scholarly field of endeavor" (Hebdige 2012, 401).

Hebdige's early statements, given their inevitable flaws, came from a supportive and constructive place. Cases of punk crossing into the academy demonstrate one of the culture's inherent contradictions. Though not necessarily "claiming illiteracy" (in fact often claiming the opposite, particularly in the case of many post-punk and hardcore bands), punk has often eschewed higher-level structuralist theses about its nature or existence. Most academics (or "punkademics") write about punk from within its fences, most often heavily embracing reflexivity while avoiding objectivity. Kevin Dunn told me, for example, how academic accounts of punk written by outsiders "are all deeply flawed because they have no understanding of the scene, leading them to make ridiculous claims or, more often, conflating 'indie rock' with punk." In his 2016 book *Global Punk: Resistance and Rebellion in Everyday Life*, Dunn cites the feminist movement Riot Grrrl as a case that avoided media interaction to avoid misrepresentation after innumerable setbacks of that nature in the late 1980s (see also Marcus 2010).

Unsurprisingly, punk's propensity to question has found an appropriate bedfellow in academia. Notable musicians, activists, and otherwise self-identifying punks have earned PhDs, while others have earned law degrees and become involved in public service. Greg Graffin, the lead singer of second-wave LA punks Bad Religion, earned his MS from UCLA and his PhD from Cornell, and serves as a lecturer in biology at UCLA. Stephen Mallinder, lead singer of Sheffield art-punks Cabaret Voltaire, earned his PhD in Australia in 2011 and recently published a chapter on critical analysis of noise (Mallinder 2013). Joe Escalante, founding member of Bad Religion's Orange County contemporaries The Vandals, earned his JD from Loyola Law School, worked for years in anti-trust litigation, and, in 2012, ran for judge in the Los Angeles Superior Court election.

Several DC scene veterans have gone into advanced degrees as well. Minor Threat guitarist Lyle Preslar earned his JD in 2007 and still practices law in New York City. More recently, Q and Not U drummer John Davis earned his master's in library science from the University of Maryland and helped found that institution's DC Punk and Hardcore Zine Archive. Without Davis' assistance, much of my archival research for this book would have been impossible. In our conversation at the UMD library, he mentioned the DC music scene's natural archival propensity, perhaps an outgrowth of Dischord's primordial function of documenting the city's nascent hardcore movement.

In words often attributed to Minutemen bassist Mike Watt, punk is whatever one makes it to be. Though punk has, like most guarded musical movements, developed tacit rulebooks for sonic or aesthetic acceptance, enforcing such ideology on simple musical performance becomes limiting and even dangerous. This kind of orthodoxy developed within various punk scenes by the end of the 1970s and alienated many founding members. Straight Edge, a DC-bred subculture of abstinence from drugs, alcohol, and promiscuous sex, is a quintessential example (see Haenfler 2004). However, punk's nature rejected (at least in corners of scenes and around the world) this orthodoxy and reemphasized individuality, originality, and iconoclasm, especially in the scene which would emerge in DC. Noted "metalologist" Deena Weinstein (2011) wrote about the obstacles to musical geography of underground music, which could be equally applied to punk and to its cousin, heavy metal:

> Mapping metal, especially its active "underground," is a messy task at best. No laws or sharpshooting border guards keep bands playing within one style, nor are there any official music guardians or academic gatekeepers enforcing the standardized usage of terminology by critics, publicists, or fans. Moreover, styles are not watertight containers: they leak, bleed into others. Musicians borrow and steal, and styles constantly evolve and transform into new styles… not even fans or critics know where to draw the lines. (p. 41)

Punk music expands this problematization even further. One of the first indigenously Parisian punk records was Métal Urbain's single "Panik." When juxtaposed with The Ramones' *Rocket to Russia* or The Sex Pistols' *Never Mind the Bollocks*, "Panik" sounds positively alien. Synthesizers and a drum machine generate noise, both effective tools for creating music yet widely taboo in many stateside punk scenes. However, Métal Urbain persisted, "kicked everybody's arse" (according to The Damned's Captain Sensible), and ultimately "claimed to create a blueprint for an authentic French punk through their recordings" (Briggs 2015, 164). After all, Métal Urbain retained the requisite loud, confrontational aesthetic through their recordings, pointedly singing in French to counteract this new iteration of British cultural imperialism. Despite factions of the Parisian punk scene adopting the raw guitar-bass-drums-screaming setup typical of early American hardcore since then, the most successful groups domestically retained their native tongue and electro elements.

Bérurier Noir, with their campy carnival-inspired stage costumes and prominent drum machine, became one of France's most successful and politically active punk bands in the 1980s and 1990s. They also refused to sing in English, and like their DC counterpart Fugazi (in attitude if not sound), they transcended punk and became what *Maximumrocknroll* referred to in 1992 as the first proper French indie rock band. They remain highly influential and respected among French punks, including many of those interviewed for this book. Many DC bands who later took their cue from electronic punk like Girls Against Boys began in earnest as side projects. Scott McCloud, who first played Paris with Soul Side in 1989 and would eventually move there in the early 2000s, found himself in this position. McCloud wanted to incorporate more synthesizers into his music, but among the DC post-punk scene in the mid-1980s, it was frowned upon. "The truth is I always wanted to play in a band with keyboards, but this was not a very 'cool' option back in the punk days" (McCloud, Email Correspondence, 4 Aug. 2016).

None of this discredits Paris or any other peripheral (to early punk) city as a viable hearth. Stylistic boundaries were not fair or applicable anywhere. Ironically, the first alleged use of the word "punk music" was on an early-1970s flyer for a performance by Suicide, a New York electronic duo of Alan Vega and Martin Rev. By the time The Ramones released their first record in 1976, keyboards had simply fallen out of fashion in the US and UK punk scenes, which quickly coalesced with major label support and sensationalist TV coverage.

Zines and Punk Circulation

Mathieu Gélézeau at home in Paris with DC-focused back issues of his zine *Positive Rage*, 2015. (Photo by the author)

> Punk music was seen as an alternative to the mainstream music industry and provided something new and liberating through its independent and 'do-it-yourself' approach.... Punk fanzines attempted to recreate the same buzz visually. (Triggs 2006, 70)

The role of fanzines in punk circulation, both in its embryonic phase and during its underground submersion and reinvention since then, cannot be overstated. Like the foundational elements of punk in the 1970s, zines were nothing new. The self-publishing tradition predates all forms of mass media. In colonial Pennsylvania, Benjamin Franklin wrote and self-published a literary magazine for psychiatric patients. Later in life, Franklin was the US minister to France and arguably more responsible than any early American statesman for forging diplomatic relations between the Continental Congress and the French crown. Across the pond, Thomas Paine's *Common Sense* followed a traditional zine archetype, becoming the catalyst for what turned into the American independence movement. In fact, small-scale self-publication was one of many industrious traits that impressed the French diplomat Alexis de Tocqueville, who wrote in *Democracy in America* (1835):

> In America, parties do not write books to combat each other's opinions, but pamphlets, which are circulated for a day with incredible rapidity, and then expire. In the midst of all these obscure productions of the human brain appear the more remarkable works of a small number of authors, whose names are, or ought to be, known to Europeans. (de Tocqueville 1978, 173)

Almost 150 years later, politics and society had profoundly changed, but the spirit of independent publishing remained relevant, especially for musicians. Where the local press' interest and network may have been limited, zinesters self-published reviews, interviews, and exposés on bands they felt deserved attention. Given how national music press was often limited in access and interest in these scenes, zine culture rose up to fill these gaps in the increasingly consolidated media (Duncombe 2008). Bands often published their own zines and sold them at shows as companion pieces to their music. Some bands during the early harDCore era never made any official recordings (or whose recordings never saw official release), so their appearances in zines were for a while the only tangible evidence that the band even existed. Insurrection, the high school band of Guy Picciotto (who would later play in Rites of Spring, One Last Wish, and Fugazi), was one example.

Zines have provided an honest, alternative perspective on an often-ignored set of contributions to music scenes. They call attention to the politics of knowledge production; because they were integral in the transmission of underground ideals, they arguably made interventions in what had been deemed as "newsworthy" or politically relevant. According to NME interviews with Tony Parsons at the time, "there is evidence of early zine exchanges between the UK punk scenes and those across continental Europe. For example, two French zines, *I Wanna Be Your Dog* and *Malheureusement*, could easily be found within the early London punk scene" (Dunn 2016b).

Long considered ephemeral, individualized statements on underground music (as records were, generations ago), zines have grown in influence and importance to methodology in understanding cultural circulation as well as public memory. Thirty years ago, zines helped circulate music scenes, and since then, they have richly contributed to the legitimization and archiving of the underground. Since 2010, the University of Maryland and the DC Public Library have both opened successful archives that were indispensable to this book. The University of Maryland's Michelle Smith Performing Arts Library opened the first zine archive collection[1] of its kind in 2014. The collection has been curated by John Davis, a Maryland alum who spent most of his life playing drums and touring with the Dischord bands Corm, Q and Not U, and Title Tracks. He also published zines as a teenager and college student, including *Held like Sound* and *Closed Captioned*. Through his expansive network, Davis has been able to secure contributions from private collections of punk fans worldwide. The UMD archive also, helpfully, includes an archive of zines from outside DC that reported on bands and activities from the scene. One example, *Alien*, based out of Grignon in southeastern France, included features on Jawbox and Nation of Ulysses in its first issue in 1994.

Recovering and reconstructing histories from the pages of zines presents many of the same challenges as gathering oral histories. Zines were self-published and wildly unregulated, so most of the reviews and interview transcripts were subject to the whims of the publisher. Most zines were produced on exceedingly low budgets, so fact-checking, editing, and even spell-proofing were luxuries. Additionally, though some of the most groundbreaking music writing has been done by teenagers, the immaturity of some zinesters pores through in the pages. Some zines which provide excellent contextual data in scene histories also include uncomfortable language or juvenile humor. In Insurrection's zine, the band posted a fake "win a date with Insurrection" contest alongside a cartoon illustration of the band performing to a crowd of screaming girls (including a prominent pair of French girls swooning over Picciotto).

---

[1] The official zine archive online catalog is available to the public online at http://digital.lib.umd.edu/archivesum/actions.DisplayEADDoc.do?source=Mdu.ead.scpa.0195.xml

Obviously, the content was created by teenage boys and predisposed to being "just as moronic as most things boys do at that age," but it provides an honest window into the psyche of the young Washingtonians who associated Frenchness with sophistication, which they sought to reflect back on the band.

Bibliography

Adams, P. C. (2007). *Atlantic reverberations: French representations of an American presidential election*. London: Ashgate.

Andersen, M., & Jenkins, M. (2001). *Dance of days: Two decades of punk in the Nation's capital*. New York: Akashic Books.

Azerrad, M. (2001). *Our band could be your life: Scenes from the American indie underground, 1981–1991*. New York: Little, Brown.

Bauman, Z. (1994). Desert spectacular. In K. Tester (Ed.), *The Flâneur* (pp. 138–161). London: Routledge.

Becker, H. S. (1967). History, culture and subjective experience: An exploration of the social bases of drug-induced experiences. *Journal of Health and Social Behavior, 8*(3), 163–176.

Bottà, G. (2009). The city that was creative and did not know. *European Journal of Cultural Studies, 12*(3), 349–365.

Bourdieu, P. (1984). *Distinction: A social critique of the judgment of taste*. Trans. Nice, Richard. Cambridge, MA: Harvard University Press.

Briggs, J. (2015). *Sounds French: Globalization, cultural communities, and pop music, 1958–1980*. Oxford: Oxford University Press.

Cathcart, R. S. (1972). New approaches to the study of movements: Defining movements rhetorically. *Western Speech, 36*(2), 82–88.

Connolly, C., Clague, L., & Cheslow, S. (1988). *Banned in DC: Photos and anecdotes from the DC punk underground* (7th ed., pp. 79–85). Washington, DC: Sun Dog Publications.

Cosgrove, D. (1989). Geography is everywhere: Culture and symbolism in human landscapes. In D. Gregory & R. Walford (Eds.), *Horizons in human geography* (pp. 118–135). New York: Barnes & Noble Books.

Crawford, S. (Writer). (2015). *Salad days: A decade of Punk in Washington, DC (1980–1990)*. New Rose Films.

De Tocqueville, A. (1978). *Democracy in America* (1835, 21st ed.). New York: Mentor.

Duncombe, S. (2008). *Notes from underground: Zines and the politics of alternative culture.* New York: Microcosm Publishing.

Dunn, K. (2016a). Email correspondence, 16 June.

Dunn, K. (2016b). *Global punk: Resistance and rebellion in everyday life.* New York: Bloomsbury.

Earles, A. (2014). *Gimme Indie Rock: 500 essential American underground rock albums 1981–1996.* Minneapolis: Voyageur Press.

Freud, C. (1959). Portrait of the beatnik. *Encounters, 12*(6), 42–46.

Gibson, C., & Connell, J. (2007). Music, tourism and the transformation of Memphis. *Tourism Geographies, 9*(2), 160–190.

Gotham, K. F. (2005). Tourism gentrification: The case of New Orleans' vieux carre (French quarter). *Urban Studies, 42*(7), 1099–1121.

Green, N. L. (2014). *The other Americans in Paris: Businessmen, countesses, wayward youth, 1880–1941.* Chicago: University of Chicago Press.

Haenfler, R. (2004). Collective identity in the straight edge movement: How diffuse movements foster commitment, encourage individualized participation, and promote cultural change. *The Sociological Quarterly, 45*(4), 785–805.

Hall, S. (1968). *The hippies: An American 'moment'.* Birmingham: Centre for Contemporary Cultural Studies, University of Birmingham.

Hall, M. M. (2016). Cold wave: French post-punk fantasies of Berlin. In M. M. Hall, S. Howes, & C. M. Shahan (Eds.), *Beyond no future: Cultures of German punk* (pp. 149–166). New York: Bloomsbury.

Hebdige, D. (1979). *Subculture: The meaning of style.* New York: Routledge.

Hebdige, D. (2012). Contemporizing 'subculture': 30 years to life. *European Journal of Cultural Studies, 15*(3), 399–424.

Hernandez-Sang, V. (2016, March 4). *"All are welcome in our band": Pan-Latino and inclusive social cohorts in Washington DC's Latin (American Popular Dance) music scene.* Paper presented at the Society for Ethnomusicology, Southeast and Caribbean Chapter Annual Meeting, San Fernando, Trinidad & Tobago.

Hoelscher, S. (2009). Landscape iconography. In N. Thrift & R. Kitchin (Eds.), *International encyclopedia of human geography* (pp. 132–139). Amsterdam: Elsevier.

Hyra, D., & Prince, S. (2016). Preface. In D. Hyra & S. Prince (Eds.), *Capital dilemma: Growth and inequality in Washington, DC* (pp. xiii–xxvi). London: Routledge.

Jacobsen, H. N. (Ed.). (1965). *A guide to the architecture of Washington, D.C.* New York: Frederick A. Praeger.

Johansson, O., & Bell, T. (Eds.). (2009). *Sound, society, and the geographies of popular music.* Farnham: Ashgate.

Knupp, R. E. (1981). A time for every purpose under heaven: Rhetorical dimensions of protest music. *Southern Speech Communication Journal, 46*(4), 377–389. https://doi.org/10.1080/10417948109372503.

Kong, L. (1995). Popular music in geographical analyses. *Progress in Human Geography, 19*(2), 183–198.

Kruse, H. (1993). Subcultural identity in alternative music culture. *Popular Music, 12*(1), 33–41.

Lee, S. (2005). Punk "noir": Anarchy in two idioms. *Yale French Studies,* 177–188.

Long, P. (2014). Popular music, psychogeography, place identity and tourism: The case of Sheffield. *Tourist Studies, 14*(1), 48–65.

Mallinder, S. (2013). Sounds incorporated: Dissonant sorties into popular music. In M. Goddard, B. Halligan, & N. Spelman (Eds.), *Resonances: Noise and contemporary music* (pp. 81–94). Los Angeles: Bloomsbury.

Marcus, S. (2010). *Girls to the front: The true story of the riot grrrl revolution.* New York: Harper Collins.

Masserman, J. H. (1967). The beatnik: Up—, down—, and off—. *Archives of General Psychiatry, 16*(3), 262–267.

McCloud, S. (2016). Email correspondence, 4 Aug.

Médioni, G. (2007). *30 Ans de Rock Français: de Telephone a Dionysos.* Paris: L'Archipel.

Morrison, T. (1999). Spider in the Snow (The Dismemberment Plan). On *Emergency & I* [Compact Disc]. Washington, DC: DeSoto Records.

Nettl, B. (2005). *The study of ethnomusicology: Thirty-one issues and concepts.* Urbana: University of Illinois Press.

Nettleford, R. M. (1970). *Mirror, mirror: Identity, race, and protest in Jamaica.* Kingston: W. Collins and Sangster.

Nevarez, L. (2013). How joy division came to sound like Manchester: Myth and ways of listening in the neoliberal city. *Journal of Popular Music Studies, 25*(1), 56–76.

Olsen, M. (1998). Everybody loves our town: Scenes, spatiality, migrancy. In T. Swiss, J. Sloop, & A. Herman (Eds.), *Mapping the beat: Popular music and contemporary theory.* London: Wiley-Blackwell.

Porcello, T. (2005). Music mediated as live in Austin: Sound technology and recording practice. In P. D. Greene & T. Porcello (Eds.), *Wired for sound: Engineering and technologies in sonic cultures* (pp. 103–117). Middletown: Wesleyan University Press.

Shernoff, A. (1975). "Master Race Rock" (The Dictators). On *Go Girl Crazy!* [LP]. New York: Sire Records.

Simpson, G. E. (1955). The Ras Tafari movement in Jamaica: A study of race and class conflict. *Social Forces, 34*(2), 167–171. https://doi.org/10.2307/2572834.

Smith, N. (1996). *The new urban frontier: Gentrification and the revanchist city.* London: Routledge.

Stokes, M. (Ed.). (1994). *Ethnicity, identity and music: The musical construction of place.* Oxford: Berg.

Taylor, H. (2001). *Circling Dixie: Contemporary southern culture through a transatlantic lens.* New Brunswick: Rutgers University Press.

Thornton, S. (1996). *Club cultures: Music, media, and subcultural capital.* Middletown: Wesleyan University Press.

Triggs, T. (2006). Scissors and glue: Punk fanzines and the creation of a DIY aesthetic. *Journal of Design History, 19*(1), 69–83.

Urry, J. (1990). *The tourist gaze: Leisure and travel in contemporary societies.* London: Sage.

Warne, C. (2013). Graphical terrorism? Bazooka, punk and leftist politics at Libération newspaper in 1970s France. *History Workshop Journal, 76,* 212–234.

Weinstein, D. (2011). The globalization of metal. In J. Wallach, H. M. Berger, & P. D. Greene (Eds.), *Metal rules the globe: Heavy metal music around the world* (pp. 34–59). Durham: Duke University Press.

Wetzel, R. (2012). *The globalization of music in history.* London: Routledge.

Capitol Crisis #1, 1980 (DC)

Mole #6, 1994 (DC)

3

A Brief History of Franco-American Circulation in the Twentieth Century

> A city does not become historic merely because it has occupied the same site for a long time. Past events make no impact on the present unless they are memorialized in history books, monuments, pageants, and solemn and jovial festivities that are recognized to be part of an ongoing tradition. An old city has a rich store of facts on which successive generations of citizens can draw to sustain and re-create their image of place. (Yi Fu Tuan 1977, 174)

It would be impossible to use any cultural circulation between France and the US, or Paris and Washington, DC, by proxy without first staging the historical basis for that exchange. These circulations flow along established channels, and this chapter aims to tell the story of those channels. No discussion of the spread of American underground culture into France would be possible without a contextual understanding of the relationship the two countries had grown through both mainstream and underground culture for generations. Paul Adams (2007) summed up the long important and lopsided appeal of America to the French in these terms. To the French, the American landscape has presented a "savage and untamed character" not present in their orderly, antiquated society. Though the relationship between the two states and their diversities of nations has

© The Author(s) 2019
T. Sonnichsen, *Capitals of Punk*, https://doi.org/10.1007/978-981-13-5968-2_3

ebbed and flowed in the past few centuries, the classic imaginaries and memetics persist.

Even the US, assuming the mantel as one of most powerful countries on Earth in the late 1940s, helping restore order in France did not change anything fundamentally. Though World War II recovery did present a sea change, laying the tracks upon which the American cultural tidal wave of rock music would roll in, it was still only the logical continuation of large-scale international cultural circulation. Popular culture had been circulating vibrantly between the US and France for centuries. Under the thumb and behind the filter of class-focused English enforcement of "high" culture, the French sought relief in a "readily available and easily enjoyed" (Gillett 1970) American culture. *Les copains*, French counterparts to the American sock-hopping teens later mythologized in *Happy Days*, were primed.

In cosmopolitan Paris, American music and film had been capturing the French imagination for as long as those industries existed. Though American film and music had travelled on the backs of individual agents before 1917, the mass movement of the Great War doughboys created the first tidal wave of American folk in Paris. By the onset of the Roaring Twenties that followed WWI's chaos, France was concerned with reestablishing order in the face of losing over a million young men in battle (Scriven et al. 1995). Fortunately, the French had the "capital of the 19th century" sitting in the heart of their nation, and Paris' cache would soon emerge as a site of unprecedented peacetime international exchange.

Though wealthy Americans in the mold of Ben Franklin had lived in Paris since before American independence, a majority of American tourism in the nineteenth century was still the province of upper classes engaging in manicured "grand tours" (see Nelson 2013). As Nancy L. Green (2014) chronicled, the Great War presented France with a new social class of Americans: the "popular" or the "folk." Legions of American servicemen, many of whom came from rural communities and had never seen a big city, fell in love with the City of Light. Some met their future significant others, and many found excuses to stick around after the War ended. Those who returned to their small towns in the rapidly industrializing US had trouble readjusting to the quiet agrarian lifestyle. Inspired by a newfound international love affair and social change in the US, Walter Donaldson, Sam Lewis, and Joe Young penned one of the most enduring pop standards of the era: "How Ya Gonna Keep 'Em Down on the Farm (After They've Seen Paree)?" Experiencing Europe, especially France, during World War I (WWI) gave black Southerners a window into the world many had not known existed, or at least never considered within their grasp. Servicemen who did return to the US soon began "the great migration," leaving their rural lifestyles in the South for the call of more tolerant, hopeful cities up North. Of course this included DC.

The movement of increasingly diverse American folk culture and popular music into Paris came at the onset of the Roaring Twenties. Black doughboys who grew up under Jim Crow laws in the American South were surprised to discover how racism, though it existed, was not so institutionalized in France. The French retained colonial holdings all over West Africa and Indochina at the time, but in that country, African Americans had found a tacit respect which had been systematically suppressed back home. By the middle of the decade, a lively African American community grew in counterpoint to the predominantly white "lost generation" of Gertrude Stein, Ernest Hemingway, Ezra Pound, John Dos Passos, and a cadre of others. African American entertainers like Josephine Baker enjoyed success and prestige in ways unimaginable in the US. Nancy Green (2014) writes:

Paris provided jobs for jazzmen. The African-American community, from down-on-their-luck trumpeters to boldly dressed "chocolate dandies" strutting along the boulevards, was largely centered on the clubs, the cabarets, and the inexpensive residential hotels of Montmartre, painted by some black artists such as Palmer Hayden and Archibald Morley. The nightclubs of Bullard, Ada Smith (known as Bricktop), and Baker formed the heart of the neighborhood: Le Grand Duc, Bricktop's, and Chez Josephine, along with Florence Jones' Chez Florence. As the French became fascinated with jazz for its sound and its meaning – liking jazz was also a way of criticizing American racism – the new music became wildly popular. This drew jazz musicians of renown to Paris and led others to take up the trumpet in order to stay on. (p. 26)

Paris, and the greater scale of France, was a destination for African American musicians throughout the jazz era. Saxophonist Sidney Bechet and violinist (and DC native) Will Marion Cook first performed in Paris in 1919 (Wetzel 2012). By the 1930s, stars like Louis Armstrong and Washingtonian Duke Ellington had traveled to Paris, finding an unprecedented respect for their form as an art, rather than raw entertainment. Dizzy Gillespie has been quoted as saying "jazz is too good for Americans" (Taylor 2001). Half a century later, their experiences would prove relatable for the second and third generations of American punk rockers.

In the mid-1970s, Parisian jazz collectors spearheaded LP reissues of American jazz and blues recordings. Henri Renaud supervised the "Aimez vous le Jazz / Do You Like Jazz" series for Columbia records, drawn from the personal collection of Dr. L. Charles Clavié. All of the releases in the series sold impressively in France (Billboard 1977). Jazz plays an indispensable role in this story of punk's circulation between DC and Paris because, like other cultural forms, it was always in the ether. Brendan Canty, who would generate a commonly accepted "rhythmic signature" of DC punk in the 1990s as the drummer for Fugazi, came largely from this genre world.

"My favorite drummer of all time is Tony Williams," Canty told the *One Week/OneBand* music blog in 2012, "[who joined Miles Davis'] band when he was 16 years old and he was just a monster. He really opened up for me what you could do with your small quiet exchanges between cymbals and hi-hats. Little things can mean a lot. He was why I started playing Gretsch drums. I was really trying to be Tony Williams."

Though DC was never considered a global watering hole for jazz like Paris, New York, or Tokyo, the District area had plenty of venues for young percussionists like Canty to catch the masters at work. Bohemian Caverns, HR-57, and other bars along the U Street corridor close to The Duke's childhood home were often places to catch live jazz, as well as Blues Alley in Georgetown.

"I was always sneaking down to Blues Alley whenever I could when I was in high school to see any of my heroes like Elvin Jones or someone like that, because you knew they weren't going to be on this earth for very long," said Canty.

The French film industry also became a platform for thick cultural circulation between that country and the US. Some of the cornerstone analyses of American film post–World War II came from André Bazin and Étienne Chaumeton. One of France's most enduring artistic movements came from this reflection on American cinema: *la nouvelle vague* (new wave). The increasing popularity of Hollywood films in France "led many post-war French critics to look for ways to understand and explain how works so clearly produced in a 'factory' could achieve such extraordinary individuality" (Kelly et al. 1995, 172–174). By the late 1950s, the post-War French fascination and deep critical analysis of American film generated a new school of French filmmaking that still captivates a global network of film scholars and enthusiasts.

Appropriately, the term "nouvelle vague/new wave" became hip to the point that the entertainment industry would eventually co-opt and contextualize it around music. Toward the end of the 1970s, music producers and fans applied the term label to the music that grew out of punk and onto the pop charts. In DC, The Slickee Boys, The Nurses, and Urban Verbs would approach this label as they separated themselves from a stagnant rock scene. Ironically, they accomplished this by turning back the clock and drawing from the original fire of rockabilly and rock 'n' roll, as was a directive aesthetic of punk. Following centuries of similar fetishization of transgressive American art forms, French rock fans responded enthusiastically.

America, Rock 'n' Roll, and the Assault on High Culture in France

American rock 'n' rollers provided templates for the first class of French rock stars like Johnny Hallyday (born Jean-Philippe Smet in 1943), whose stage name was a deliberate nod at extant templates for musicians like Johnnie Ray, Elvis Presley, and Gene Vincent. Though Hallyday would not break through to Anglophones as his contemporaries Serge Gainsbourg and Jacques Brel later managed, he remains a household name throughout the French-speaking world and a valid starting point in the conversation about nationalism in French pop culture (Looseley 2005; Briggs 2015). Hallyday's career, which has continued through at least one "retirement" and well into his 70s, has weathered the waves of public acceptance of rock 'n' roll in mainstream French culture. Charting Hallyday's career could be read as another case study of French popular music falling a few years behind the curve, as Parisian hardcore did when it coalesced out of punk in the mid-1980s.

"France is not a very rock 'n' roll country," said Thrashington DC vocalist Fabrice Le Roux. "We don't have a rock tradition. I don't see France as [that]. In my opinion, France is always lagging behind … Even when bands from the US come to France to tour [today], they only come here to cross the country to get from Germany to Spain. In my opinion, Germany [has] more [of a] rock tradition."

When Hallyday emerged as France's preeminent rock star in the early 1960s, he represented a threat to the establishment similar to how Elvis Presley had terrified middle America five years prior. By this point, the conservative backlash to rock 'n' roll had died down somewhat. Elvis Presley had entered military service and other standouts like Buddy Holly and Ritchie Valens had died tragically. Others, like Johnnie Ray, faded from the limelight, though he remained successful outside the US. Johnny's arrival "was feared by the cultural and political establishment everywhere as a dangerous influence on youth, and in Gaullist France particularly, as a vehicle of US cultural imperialism" (Looseley 2005, 200). Of course, these attitudes would change. The establishment's acceptance and celebration of Hallyday reached a point that, in 2003, state officials and some of the most powerful French businesspeople reserved more than 800 seats at his sixtieth birthday party in Paris.

The American infiltration and French embrace of rock 'n' roll culture (which would shift into "rock" in its own context, as it had in both the UK and the US previously) also pinpoints one of several notable moments which compromised the gap between "high" and "popular" culture. For most of modern history, established "culture" (music, theater, literature) had largely been the province of the elites—those who controlled its publication in whatever formats existed. Prior to the Enlightenment era following the Renaissance, the elites were the only ones who could even read or write through most of Europe. The term "popular" had a negative connotation until the late eighteenth century, when meanings shifted from "low" or "vulgar" to more "widespread" or "accepted," gradually accruing more positive associations over the course of the nineteenth century. As Anahid Kassabian (1999) puts it, "this history is important because of the meaning of the term shifts from embracing the perspective of an elite class that looked down its collective nose at the common people, to celebrating – and remaking – what those common people valued" (p. 114).

Sociologist Howard Becker offers the classic model for understanding the ability of culture to create social formations through what he called "art worlds." The interaction between artistic values and artistic production establishes a lexicon for a social group to speak to one another, to share ideas. Although Becker focused on the production of high art, Simon Frith (1996) illustrates how Becker's ideas are just as applicable to popular music; much like high art, the value and meaning of popular music are based on a combination of musical expression, critical mediation, and an audience reception (Briggs 2015).

The eventual contributions of DC punk to the canon, then, come to mind here. Not a hardcore band per se but unquestionably a product of hardcore, Fugazi were, perhaps more than most of their contemporaries, responsible for elevating punk music into the domain of high art. Their output may not have the same esteem among music historians as that of Bach, Berlioz, or Beethoven, but within the realm of popular music, they have helped establish punk as intelligible to intellectual aesthetics while also helping shed some of the genre's stigma among critics.

This has all happened commensurate with the gap between popular and "high" culture closing. The wealthy elite, while physically isolated, rarely avoid popular culture given the contemporary media proliferation. Meanwhile, the population of westerners who came of age prior to the rise of rock 'n' roll is drastically thinning. The youngest person to remember popular music before Elvis Presley is at least 70 years old. However, rather than separating rock music from the realm of high culture, this progression has redefined the idea, perhaps regimenting and segmenting grades within popular culture. Pierre Bourdieu, the influential French social theorist, wrote in his cornerstone *Distinction* (1984) that

[t]he embodied cultural capital of the previous generations functions as a sort of advance (both a head-start and a credit) which, by providing from the outset the example of culture incarnated in familiar models, enables the newcomer to start acquiring the basic elements of the legitimate culture. (p. 71)

In other words, "legitimate" culture has a shelf life, and it cycles out. This has been proven true on both sides of the Atlantic. As Johnny Hallyday exemplifies, "the standard Bourdieusian view – that classical, or 'bourgeois,' culture is socially constructed as legitimate in order for its elite proponents to be demarcated from the mass – has been overtaken by events in the last quarter-century, because popular cultures themselves have undergone a legitimation process in their own right" (Looseley 2005, 201).

Punk Comes to DC with French Assistance

Though DC punk is most often associated with Dischord Records and a handful of bands praised as seminal by underground music aficionados, the actual city's legacy of unfriendliness to the performing arts has slimmed the volume of documented material, yet aided its archival accessibility. Through the early 1980s, and even until scene denizens Dave Grohl and Dante Ferrando opened The Black Cat up the street from HR-57 in 1993, the epicenter of the Washingtonian punk scene was the 9:30 Club. Other clubs like the Bayou hosted some left-of-center rock acts, but only 9:30 catered to the purposefully confrontational and experimental. An ad for Nightclub 9:30 placed in the March 1981 issue of the zine *Dischords* features a black-and-white photo of a young woman in a mask, mixing chemicals and creating smoke amid a batch of lab equipment.

For years, the nightclub operated in the old Atlantic Building at 930 F Street NW. The surrounding downtown was in a notoriously blighted condition and featured little else that brought anyone upwardly mobile to the area after dark. The Atlantic Building's neighbor had been, between 1918 and 1968, the Georgian Revival Metropolitan Theater, one of the first three movie palaces to open in downtown, along with the Rialto and the Palace (Goode 2003). The Metropolitan stuck it out through the post-War downtown slump until it shuttered after the assassination of Martin Luther King, Jr., and the subsequent riots in April 1968.

The 9:30 Club hosted one of Minor Threat's final performances in 1983, along with a host of other legendary shows by local and touring bands. While other performance spaces began to fade, particularly with the District's 1995 funding pull, 9:30 managed to weather the storm. In 1996, the club relocated to 815 V Street, a roomier space on the eastern end of the gentrifying U Street corridor. The Atlantic Building, and the last ghosts of the original 9:30 Club, came down in 2000 (Goode 2003).

The city had a notoriously paltry club scene in the 1970s; there were few nationally visible homegrown bands, and the city in turn did not attract too many mid-size acts. The Capitol Center, which opened in suburban Prince George's County, Maryland, in 1973 to accommodate the Bullets basketball team, provided a venue for major stadium acts. The building's first years saw visits from the Allman Brothers, the Who, Frank Sinatra, and Elvis Presley on his final tour. As the arena-rock era coalesced in the late-1970s with appearances from Van Halen and Ted Nugent, the Capitol Center concerts became synonymous with the putrid excesses to which the city's coming-of-age harDCore generation took exception. On May 31, 1986, local A/V nerds John Heyn and Jeff Krulik brought video cameras to the Capitol Center parking lot to interview metalheads prior to a Judas Priest concert. They unintentionally immortalized this highly intoxicated subculture in their short film *Heavy Metal Parking Lot*. Underground circulation of the movie via VHS traders over the following two decades would afford Krulik and Heyn cult status among aficionados. The shoestring documentary became a favorite on Nirvana's tour bus in the early 1990s, and Dave Grohl has been quoted calling it "the greatest film ever made about rock n' roll."

Looking at DC's marginal pop culture identity at the time, the multigenerational appeal of *Heavy Metal Parking Lot* makes sense. At the height of MTV's power, the bands were becoming larger than life while few people were turning cameras on the fans. Few DC artists had landed major record deals over the past decade. The glam-metal band Angel signed to Casablanca after being discovered by KISS, but their success was fleeting. The Urban Verbs, led by Roddy Frantz (brother of Talking Head Chris Frantz), managed a two-album deal with Warner Bros. Records in 1978 on the heels of several trips to play CBGB's in New York, but their sound hardly transcended what other new wave bands were playing to bigger audiences at the time.

The Slickee Boys, regarded as DC's first new wave band, however, remained firmly local. The biggest label they ever signed with was Twin/Tone, a Midwestern label best known as the incubator of The Replacements in the early 1980s. Though Twin/Tone was not exclusively a punk label, it became emblematic of the wave of "DIY punk labels [that] sprung up globally partly because global capitalism's attempt to profit off of passive consumers actually led to the development of a vibrant independent, anti-capitalist DIY punk culture" (Dunn 2016, 136).

Founding guitarist Martin "Kim" Kane, who spent his childhood in Korea, named the band after a youth movement he encountered overseas. Including guitarist Marshall Keith and vocalist Martha Hull, the group recorded the covers-heavy *Hot and Cool* EP in 1976. Given how barely any major-label focus rested on DC at the time, Kane and Keith created Dacoit Records in order to release it. The following year, Half Japanese (aka ½ Japanese), an outsider group from remote Uniontown, Maryland, led by brothers Jad and David Fair, would do the same with their label 50,000,000,000,000,000,000,000,000 Watts Records. Though they were geographically separated from the District area early on, DC was the closest accepting music scene and city at the time and they came to be widely regarded as quirky pillars of that city's boundary-free urban ethos.

It was in the stagflated 1970s that people well outside of the new wave music scene became susceptible to the seeds that the DC atmosphere and community were planting. A nine-year-old French transplant named Fabrice Laureau spent afternoons teaching himself and his six-year-old brother Nico how to skateboard. Their father worked at the French embassy and they attended the French school in Bethesda, enjoying a bucolic, uniquely international childhood.

"Even if we only spent a part of our childhood in DC, these were important and influential years," said Nico. "Since we kept great memories of this period, it definitely brought a possibility of a strong identification."

Between 1977 and 1980, the Laureau family became involved with progressive social life in the US capital, including the Longest Walk protest and other events supporting the American Indian Movement. Though Fab and Nico were too young to be privy to the new wave scene budding downtown, they would return to DC intermittently over the following decade. Back in Paris as young adults in the 1990s, they would add a valuable building block in the circulation of punk and art (namely, art-punk) between the two cities.

An ostensible lack of boundaries (or outside recognition at the time) owed much to the city's relative invisibility, but it did not necessarily mean that DC cohered as a strong scene for that generation of new wave bands. A 1979 issue of the zine *Descenes* included "a family tree of the D.C. Underground" at the time, which included The Slickee Boys, Mark Hoback Band, and six iterations of the group White Boy. The Slickee Boys produced a few issues of their own zine intermittently, sometimes merely to document their recent activities for fan club members. One 1985 issue of *Slickzine* gave Kane the chance to print his official history of the band, as well as a platform to voice concerns over their international distribution at the hands of Twin/Tone Records.

Though they predated the local bands who would initiate the diffusion of the DC scene into London and Paris in the early 1980s, The Slickee Boys would eventually reap benefits and downfalls of the growing cultural circulation in equal measure. The group signed with Twin/Tone to release their 1983 full-length LP *Cybernetic Dreams of Pi*, and drummer Dan Palenski engineered a deal for New Rose Records to release their lead single "When I Go to the Beach" in Europe. JD Martignon, the French owner of Midnight Records, a shop across the street from New York's infamous Chelsea Hotel, got Patrick Mathé of New Rose in touch with him. Shortly thereafter, Mathé licensed the full LP from Twin/Tone. Palenski remembers receiving royalties for the "Beach" single in 1984, but is unsure if royalties came for any releases after that.

Despite these hurdles, The Slickee Boys' career demonstrated how French and American artists shared a passionate respect and curiosity for one another regardless of the city of origin. According to Palenski in an email to Kane, he was the one who talked The Slickee Boys into covering "Death Lane," a track by early Rouen punk upstarts The Dogs. George

Budnovitch, a friend of the Slickees who had been pen pals with Dogs singer Dominique Laboubée, had given Palenski a copy of that band's 1982 album *Too Much Class for the Neighborhood*. Apparently, The Dogs got a huge boost of prestige in France when word got around that an American band had covered one of their songs when The Slickee Boys included it on *Uh Oh No Breaks* in 1985. By that point, members from the Slickees' local circles (even those outside of punk) were also openly courting interest from French labels and reciprocating that fascination:

> Those wanky Wanktones – the Slickees favorite Maryland hillbillys, may just be signing (seriously) an LP deal with BIG BEAT RECORDS of France. Apparently though, it's Bo Link who's holding up the signing until the French company gets him an autographed picture of Johnny Hallyday (the French Elvis) and a date with Bridgit Bardot!!!! (*Slickzine* 1985)

HarDCore Takes Shape and Dischord Records Begins

Jean-François Moulard (aka Maz) founded the punk label Meantime in provincial St. Etienne in 1995 and has since become an enigmatic figure to people connected to punk all over the country. A few years ago, he shared some thoughts with me about what Dischord Records means to him, incorporating both geography and *The Situationist International*, the radical 1960s-era organization of intellectuals in Western Europe:

"The thing that was important in the 80's and 90's was that Dischord Records proved that kids could promote a scene that they created, and give it an exposure worldwide without doing concessions," wrote Maz. "Keeping it genuine and out of the 'société du spectacle' (as said *[Guy]* Debord) and still being able to have a large audience. And yes, 'putting DC on the map'! And giving an identity to what was going out of Washington in the prospect."

Rather than musical style or aesthetic as a guiding principle on what constituted a "DC sound," geography has remained the most viable component in drawing any theoretical lines around punk bands and their scene. With some notable exceptions, all of the bands most closely aligned with global imaginaries of DC, harDCore, and post-hardcore released music on Dischord. In an odd way, the label's directive of documenting the musical goings-on in the DC underground mimics an imperative of folk festivals and other movements that are "embedded in modernity… created in oppositional discourse against the commercial and the cosmopolitan, favoring the folk and the local" (Regis and Walton 2008, 409).

Bad Brains are widely considered the inventors of the hardcore punk style, and for many they are considered the best band to ever play in that style. A major reason for this lies in their origin as a technically proficient jazz fusion quartet called Mind Power. Bassist Daryl Jennifer, guitarist Dr. Know (Gary Miller), drummer Earl Hudson, and his brother vocalist Paul (H.R.) had grown up together in the city's predominantly black Southeast quadrant. They relocated to New York at the beginning of the 1980s (as chronicled in their song "Banned in DC") and they did manage some marginal mainstream success in the latter part of the decade, shifting to a more radio friendly metal-based sound. In 2016, they landed the most mainstream of musical recognition: they were nominated for induction into the Rock & Roll Hall of Fame. The local press reflected on the band's legacy:

> Even within its own genre, Bad Brains was an act unlike any other, pushing the idea of what punk rock and hardcore music could be, while breaking down racial barriers. The band fused elements of reggae, metal, R&B and funk. In that sense, Bad Brains became a bridge for from artists like The Clash and Sex Pistols to the likes of Jane's Addiction, Living Colour, Rage Against the Machine, Sublime and Soundgarden. (Smith 2016b)

"In 1979-80, not only did we have this great local band, we actually had the greatest band in the world playing in Washington," Ian MacKaye told WTOP Radio in light of the band's nomination. Many formative bands in the hardcore subgenre, including San Francisco's Dead Kennedys and Vancouver's DOA, had started playing shows in 1978. HarDCore as a phenomenon to Washington may have eventually happened if Mind Power hadn't changed their name to a Ramones song and taken a chance on punk rock, but the cultural history of the subgenre would be radically different. The band's debut single "Pay to Cum" (1979) became a lightning bolt (more on that symbolism later) that changed the course of music in DC.

"I remember hearing 'Pay to Cum' and just being attacked with it," said Shudder to Think singer Craig Wedren. "It was the rapid-fire singing. The rapid-fire playing. It wasn't metal, it wasn't punk, it was melodic, it had so many totally revolutionary [elements]... [Bad Brains] made all of our chops so much better."

For the new generation of punks coming up, an embrace of Bad Brains served as a similar rejection of the racism ingrained in their parents' politics and social landscapes that the original French embrace of jazz had been. Many of them wound up hanging out in whatever communal spaces the band could set up and perform in. Madam's Organ, a Corcoran Art students' collective located at 2318 18th Street NW, provided one of few venues that outsider bands could play. Because it was a community center at the heart of pre-gentrification Adams-Morgan, kids of various ethnicities as young as ten could be seen in the audience on any given night.

This underground movement quickly caught a spark with a group of students at Woodrow Wilson High School up in Tenleytown. At the time, the school was approximately 75% African American, and the students enjoyed relatively little oversight by the administration. Gregor Kalas was one member of the Class of 1981 who would become a recognized professor of classic architecture (influenced by coming up in a city known for landmarks built in the style). It was easy at the time for students to, for lack of a better term, make their own culture.

"Owning your own culture was a big concept for us," he fondly recalled, citing an atmosphere of anger among that generation of urban youth, mostly directed at their parents' generation for blighting the inner cities and economic stagflation. Wilson also operated fairly liberally considering the "prison-like" atmosphere at Alice Deal Middle School on the opposite side of Fort Reno Park. Deal was often the first setting where white kids from Northwest would mix with the black kids from across Rock Creek Park. Considering the racial antipathy that large-scale rioting had stoked a decade prior, the mixed-race environment took some getting used to. Symbolically, the Fort Reno free summer concert series, still held annually and many of the bands included in this story have played, started there in 1968 as a balm for a riot-scarred city.

Though they went to school at DC's highest elevation in Tenleytown, most of the kids who got into punk worked and hung out down Wisconsin Avenue in Georgetown. Four of the Georgetown Punks, who called themselves The Teen Idles (singer Nathan Strejcek, bassist Ian MacKaye, guitarist Geordie Grindle, and drummer Jeff Nelson) were unwittingly helping coalesce one major international movement while participating in another. Musically unsophisticated, the band wrote a handful of originals about their outlook and hobbies, played at Bad Brains' blinding speed. They padded their set lists with hyperactive covers of 1960s pop songs, many of "which were so vital to the success of the *copains* [French teenyboppers] as translations in the 1960s" (Briggs 2015, 158). Teen Idles opened their first gig with a chaotic rendition of the Contours' 1962 hit "Do You Love Me?" Within a year, three more foundational harD-Core bands—State of Alert (SOA), The Untouchables, and Minor Threat—recorded covers of the Monkees' 1966 hit "(I'm Not Your) Steppin' Stone," which had been previously covered by the Sex Pistols. The song would also later be covered by the French punk band Les Thugs and over a dozen other bands worldwide.

Barely out of Wilson High School when college and interpersonal conflicts spelled the end of their band, the Teen Idles decided to spend the little money they had saved to make a record. The result was a 7-inch EP, *Minor Disturbance* (Discord #1). Skip Groff of Yesterday and Today Records in Rockville, Maryland, agreed to record them on his suburban setup. The fuzzy, spastic EP contained eight songs, almost all of which clocked in under one minute. As the story continues, the band folded, glued, and packaged the records themselves. Strejcek accidentally melted dozens of copies on a radiator, thereby accelerating his exit from the Dischord web as well as making the records "even more of a collectors' item" (Andersen and Jenkins 2001).

No major label, even punk and new wave–friendly concerns like Sire or Stiff, had a pack of teenagers from Washington, DC, anywhere near their radar. Even if they had, punk music had fallen from abject marketability, and the idea that hardcore had any commercial potential was unthinkable. For these bands, doing it themselves was a necessity, not a choice. Minor Threat, the band that MacKaye and Nelson would form with Georgetown prep students Brian Baker (bass) and Lyle Preslar (guitar), toured North America, building success and esteem for their still-nascent label. At least once, the band sold out of a release solely off of pre-orders, a testament to their niche popularity. Meanwhile, Dischord issued the first releases by DC-area hardcore bands like SOA (featuring a young Henry Garfield, before he changed his name to Rollins and moved to California), Void, Faith, Government Issue, Marginal Man, and the band that would eventually beat all others to play in France, Scream.

Philippe Roizès hanging out at Scream practice behind bassist Skeeter Thompson, DC, Summer 1987

By 1983, when Minor Threat broke up, many of the fastest and loudest bands had already split, and many of the local musicians got tired with breakneck rhythms and unintelligible verse. By 1985, many of the same musicians from the first generation of harDCore mixed with kids from the second generation. The "older" generation, who oversaw the gelling of the underground scene into hardcore, splintered into varying factions of slower, more heart-on-sleeve groups. At the time, there was no frame of reference for what constituted "generations" in the underground scene. Each generation lasted roughly four or five years, considering how many of the most active players on the scene only stayed together for one or two at most.

DC's ostensible detachment from the music media allowed the first generation of hardcore bands to diversify and release music however they saw fit. By 1985, several new bands were writing slower, more confessional songs that many zine writers reacted to by labeling the style "emotional hardcore," which slowly became "emo-core," soon thereafter, "emo." Most of those labeled "emo," especially MacKaye, then fronting the short-lived Embrace, found the classification ridiculous, but it was out there.

Shortly after Insurrection broke up, Guy Picciotto cofounded Rites of Spring, named for Igor Stravinsky's landmark work *Le Sacre du Printemps* whose 1913 Paris premiere combined dissonant harmonies with primal subject matter from Russian folklore and sparked a legendary uprising from the audience at Théâtre des Champs-Élysées. Punks in the DC scene were unintentionally writing a new chapter in a quintessentially Parisian tradition of musical rebellion. Other foundational emo bands like Embrace, Dag Nasty, and Gray Matter followed suit over what came to be known as "Revolution Summer" 1985. They frequently collaborated with Positive Force, a local activism group led by Mark Andersen, to arrange benefit shows with the overarching goal of helping a struggling society from the ground of its seat of power. Punk activism, which became the law of the land around this time, would open the door for involvement for a variety of French punks who would visit in the city over the next few years.

Though 1985 may have been the year that DC punk came of age and fully embraced the "think global, act local" ethos, 1986 was the year it officially took its business internationally. In late March 1986, a few months before Scream played Paris with Sherwood Pogo, Ian MacKaye and Jeff Nelson flew to London. Their goal was to secure a distribution deal with John Loder at Southern Recordings, effectively to satisfy fan requests for Minor Threat records in the UK. At the time, neither of them was actively playing in bands, so they had a brief window to focus on curating the Dischord back catalog. While there, MacKaye and Nelson added to the catalog, laying down two tracks as Egg Hunt ("All Fall Down" backed with "Me and You"), their final recording together.

Though MacKaye would not visit France for the first time until touring with Fugazi two years later, he remembers "a guy named Pierre, that worked for Southern, a French guy [who] actually drove a Citroen" singing backups on the Egg Hunt recording. He was largely unaware of the impact that his music was having in France, but that would change profoundly over the following three years. MacKaye and Nelson struck a deal with Loder that moved the international manufacture of his label's releases to a French factory that Southern had contracted out. For decades, a majority of Dischord releases would read MADE IN FRANCE, permanently connecting the two in the global imaginaries of punk fans worldwide.

The French production of Dischord's vinyl bears an interesting story. The deal that MacKaye and Nelson struck with Southern worked out very well, very quickly. An enthusiastic and well-connected John Loder went out seeking pressing plants in mainland Europe, and he made a deal with the British company Mayking, who contracted with MPO (Moulages et Plastiques de l'Ouest), a plant in northwestern France, approximately 100 km northwest of Le Mans. The first batches of Dischord albums did not include "Made in France" labels, an infraction to which American customs agents quickly put their foot down.

"Back then, me and Jeff, when we would get a shipment from England, we would have to go out to the [Dulles] airport and actually process it," related MacKaye. "So we would have to go do all the documentation, do all the paperwork, work with the customs people, it was *insane*. We would go out there with thousands of 'Made in France' stickers and just be sticking them on. Which sucked. So we just started printing them on, which started causing problems. In record stores, you were charged more for the import records. Stores were charging more for the French pressings of the records, which was why we went so bold with the *This Record $5 PPD from Dischord Records* [text]. I had never been to France and never met any of the people over in France [at that point], it was just the way these records were labeled, so suddenly it was like we had some French connection."

Since then, Dischord has never stopped releasing music, and more crucially from a geographic standpoint, never released any music by musicians outside of the DC region. In the past two decades, the label has expanded to include some Baltimore acts, most of whom include musicians with a DC pedigree like J. Robbins. MacKaye and Nelson never felt the need to expand into other scenes, as their city had more bands to record than they could afford. Ultimately, their sustainability gamble paid off. Dischord's four million records sold are nearly inconceivable for any independent label that has never scored a hit single or chart-topping record. In fact, *Red Medicine*, Fugazi's 1995 album, peaked at 126 on the *Billboard* 200, thereby making it the highest-charting Dischord release. Fugazi's final album, 2001's *The Argument*, reached #1 on the Independent Charts, but only #151 on the *Billboard* 200, where it spent one week in early November 2001. Ironically, the #1 album that week was entitled *God Bless America*, a patriotic 9/11 tribute on Columbia Records that featured no artists from DC. These sales, along with a disproportionate amount of media coverage of Dischord and many punk fans around the world holding Dischord up as an ideal, have made the label into a monolith. This under-serves a large grouping of DC punk artists who never released anything on that label (including Bad Brains) and continue operating outside of its orbit.

In 2016, when Dischord uploaded digital files of their whole catalog to Bandcamp.com, that site's editors organized a roundtable of musicians and players in the label's history. They posed a question about the inextricable connection between Dischord and Washington, DC, geography, which received animated responses.

[Bandcamp Managing Editor / Former DC Punk Jes] Skolnik: I think that's the nature of all local punk scenes, really, but Dischord is so community-minded, and there are so many references to other local bands, political movements, monuments, and places throughout the entire catalog, both visually and musically. And I learned the ethics of being part of the community and the idea of giving back from Dischord and Positive Force. You didn't just have a show in a church basement, you did a benefit for the needle exchange that that church hosted on the weekends. I always think about something I've heard both Cynthia Connolly and Ian MacKaye talk about: that DC didn't have a larger punk scene, so they had to make their own. There's something fiercely provincial about it that I honestly like a lot. DC doesn't get enough credit, ever.

[Author and Northern VA native Joe] Gross: It was massively important to me. The idea that all of this stuff was near me—I felt a sense of ownership, of sorts. I adored its regional focus. Dischord was like a folk label in that way, and I really dug that. To this day, I have a bit of a knee-jerk, I-will-give-this-a-spin reaction to bands from anywhere from Richmond to Baltimore, and Dischord is a huge part of the reason why.

[Smart Went Crazy/Beauty Pill head Chad] Clark: I think I remember talking to Ian about the label's focus on regionality, and I think he said that he always expected other labels to conduct themselves that way. Like, it was a natural characteristic of independent labels to reflect where they came from. That makes sense to me.

[Swiz / Bluetip / Red Hare Guitarist Jason] Farrell: There were many other local labels representing their scenes. Dischord followed that same idea, but did it quite well. None lasted as long.

These public conflations of the label's down-to-earth practices with the city of DC itself had long since taken root for their international fans as well. A generation of French punk fans who came of age in the 1990s, catching Fugazi on their European tours and getting into an increasingly mobile set of DC bands, understood the connection.

"I think we were all very inspired by the indie music scene in DC – and we were trying to do something similar in our own way," said longtime fan Noémie Ventura Rimmer. "I think that Dischord Records showed us that it was possible to just do your own thing and be self-reliant without compromising your work for the sake of marketing. They are/were very inspiring folks. They represent an ideal, the embodiment of punk values. They walk the talk, it's rare... They have been standing for a long time, even after the Nirvana phase back in the nineties – when all these US indie labels back then were being bought by big record companies."

Punk Labels and Urban Ethos

The artisan readily understands these passions, for he himself partakes in them: in an aristocracy, he would seek to sell his workmanship at a high price to the few; he now conceives that the more expeditious way of getting rich is to sell them at a low price to all. (Alexis de Tocqueville, 1840 (1978, 170))

Few labels carry a greater attachment to the urban landscape of any city as Dischord does to Washington, DC. Ian MacKaye has repeatedly gone on record to say that Dischord's purpose was to document that scene.

"I like the idea that when it's all over – and it will all be over – that someone will say, 'Yeah, Dischord was from D.C.,'" he told *Anti-Matter* zine in 2007. "That's just so cool. Not only does it create an illusion, but it actually creates reality of a community. It actually creates a scene. It's no secret that Washington had a cool music scene or still has a fairly cool music scene, but part of what makes a scene is someone there to document it."

Independent record labels like Dischord have become icons of circulation of local and regional identities as much as the music they released. Many labels have kept their releases exclusive to bands from their hometowns. This was often as much a question of sustainability as it was hometown pride and ethics. For example, MacKaye has cited his genuine desire to keep bands on his label close to home and easily accessible as much as the handy excuse to politely reject overzealous artists from outside the Mid-Atlantic who send in demos. Many boutique labels around the world, like Meantime, have modeled themselves off of Dischord, which itself was modeled after Dangerhouse Records in Los Angeles. Thereby, as with music being an irreplaceable representation of place, record labels have also become conduits for circulation of punk culture and practice.

"There is generally speaking in music fans a strong attachment to locating the artists or labels they like," wrote the Paris-based electronic singer and English literature scholar Laurence Estanove in 2015. "This has a dual implication: on the one hand it may tend to individualize and demystify the artists, creating a form of fantasized proximity with them, granting them a concrete, geographically rooted existence. But it also definitely works at mythologizing both the label that gives the artists a collective (supposedly) coherent existence, and the city itself."

Numerous labels demonstrate this phenomenon on a global scale, especially in North America and Europe. Sub Pop, the label that most famously signed Nirvana along with many bands integral to the grunge explosion of the early 1990s, remains active and based in Seattle. Sub Pop bands, at the time mostly home-grown groups who shared a murky punk aesthetic, dominated the discourse on what would coalesce into "the Seattle sound" (see Bell 1998). The label expanded its focus during the 1990s, signing Midwesterners like The Afghan Whigs and, later, the successful New Mexico indie group The Shins. However, Sub Pop remains firmly intact as integral to the urban ethos of Seattle: creative, bleak, and noisy.

The Chapel Hill–based Merge Records earned a following with its founding band Superchunk and North Carolina contemporaries like Neutral Milk Hotel and The Mountain Goats through the 1990s, and eventually earned surprising mainstream success by signing the Montreal-based Arcade Fire in 2004. Merge is arguably one of the few indie labels that transcended its hometown, but its founders continue operations in North Carolina. Similarly, No Idea Records, which began as a fanzine in Gainesville, Florida, in 1985, grew into an impressive documentation of activity in the swampy land-grant college town. The label's earliest releases were predominantly Floridian, some of whom became commercially successful in the 1990s (Less Than Jake, Hot Water Music) and 2000s (Against Me!) and left the label. Though, like Merge and Sub Pop both, No Idea's roster expanded nationally very quickly, and the label remains based in Gainesville and reinvests in that city's musical legacy through the annual fest every late October.

Other successful American indie labels benefitted from their locations in major entertainment centers. Epitaph Records was founded in 1980 by members of the Los Angeles punk band Bad Religion (who former Minor Threat/Dag Nasty bassist Brian Baker would join in 1994). The label struck mainstream success when The Offspring, an Orange County quartet on their roster, scored two smash hits in 1994 with "Come Out and Play" and "Self-Esteem." Other American bands who capitalized on the well-marketed mainstream embrace of punk rock that year, including the Bay Area bands NOFX and Rancid, sold many records for Epitaph but left to start their own labels (Fat Wreck Chords and Hellcat, respectively).

Despite the cultural geographical contrasts between Northern and Southern California, labels like these successfully commodified and sold a proscribed "California" aesthetic around the world. Perhaps most significant for the DC punks (and many Parisians as well) was SST Records, a Long Beach, California-based label started by Black Flag guitarist Greg Ginn in 1977. SST, maintaining operations in the Los Angeles area for decades, released music by many of the bands with whom Minor Threat would form a kinship in the early 1980s, including Descendents, Minutemen, Meat Puppets, and Hüsker Dü. Poor financial leadership of SST would eventually drive many of the bands that outlasted Black Flag away from the label acrimoniously (see Azerrad 2001), but SST's catalog remains among the most venerated of that decade.

Green Day, the band whose mainstream success has had perhaps the greatest impact on punk's globalization (see Wallach 2008), became the guiding force and ultimately the downfall of Lookout Records. Larry Livermore technically began the label in the mountain communities between San Francisco and Sacramento in 1986 to release albums by his trio The Lookouts, but the breakout success of their late-1980s signings Green Day, The Mr. T Experience, and Operation Ivy moved the label's geography firmly to the East Bay. Ultimately, poor accounting practices, unsustainable distribution agreements, and Green Day needing to pull their back catalog resulted in the label going bankrupt and eventually shuttering in 2012. Despite the label's unsustainable success and consequential failure, which Livermore chronicled in his 2015 book *How to Ru(i)n a Record Label*, Lookout demonstrates how firmly a legacy of place and urban ethos cling to a record label well past its life cycle.

Another Bay Area label that did not enjoy as much mainstream success was Alternative Tentacles. Dead Kennedys vocalist Jello Biafra founded the label in 1979 out of the same necessity that sprouted labels like Dischord and SST, but his aim was more iconoclastic. His trademark far-left politics and confrontational attitude necessarily skirted the capitalism-driven pop charts, which still attached the label firmly to the volatile urban ethos of San Francisco. As Michael Stewart Foley (2015) describes at length, the Dead Kennedys' vicious attacks on conformity, Ronald Reagan's encroaching "Morning in America," and atrocities like "Chemical Warfare" captured the zeitgeist of an unstable city on the verge. That the band's first gigs happened within a month of Dan Brown murdering Harvey Milk and George Moscone was not a coincidence; the simultaneous genesis of hardcore punk in DC was a similar confluence of social and political factors that the musicians behind it could not help but live among. In fact, the Teen Idles' summer 1980 trip to San Francisco and Los Angeles exposed the kids to California punk labels like Alternative Tentacles, SST, and Dangerhouse, all of which inspired them to begin Dischord back home.

Similarly, in the UK, local indie labels sprouted up in reaction to a London-dominated music industry that tended to overlook provincial scenes. Postcard Records, founded in Glasgow by the enigmatic Alan Horne in 1980, gave start to underdog post-punk groups like Orange Juice, Aztec Camera, and Josef K. Its similarity to Dischord would be paramount had Horne not seized operations in 1981. Regardless, Postcard's kinship with Dischord merely begins at the locals-only approach and their 1980 origin. Horne and MacKaye were not interacting at the time, but both lived in cities they felt were cruelly overlooked and were reacting to that injustice. Decades later, the Glasgow and DC addresses associated with their labels remain pilgrimage sites for obsessive fans.

Such was the case for the Bristol label Sarah, which focused on releasing 7-inch records by bands often described as twee-pop, a delicate late-1980s substrata of UK indie and ostensible counterpart to emo. Most of the original Sarah releases were intentionally and necessarily limited in pressing, so many of these items are among the most prized pieces for collectors of British indie music. Similar to Horne's treatment of Postcard, the label founders Clare Wadd and Matt Haynes purposefully put their label and brand to bed. They thereby enhanced the mythology which surrounds the label and their iconic set of artists which included The Field Mice, The Orchids, and Another Sunny Day. Critics have cited the latter's 1988 single "I'm in Love with a Girl Who Doesn't Know I Exist" as summation of both the twee-pop movement and what the then-nascent emo movement would become in the 1990s. Though not all of Sarah's artists were from Bristol, the label has illuminated that city in the mental map of the UK for indie pop fans worldwide.

As in many departments, French contributions to this subcultural phenomenon have not gone as widely recognized. By the time that Meantime began, Fugazi and other Dischord groups (Scream, Shudder to Think, the Fire Party) had played multiple gigs throughout France and their label was beginning to intrigue scene enthusiasts. Maz started Meantime Records in order to release the first record by his band Sixpack. Thereafter, Maz decided to maintain Meantime as a Dischord-like experiment, only releasing material by bands in St. Etienne.

"It was important from the start to release mainly material of local bands," Maz said. "The label Bonanza Recordings, run by my mate Frank, was already doing just this, but there [were] a lot of bands in St. Etienne then, so I [wanted to] help... I've always been into doing things in my neighborhood first, it's almost a political thing, really."

Maz did the same thing concurrently via the Meantime zine, interviewing local bands and trying hard to "put St. Etienne on the map." Over the turn of the century, Meantime started to collaborate with bands from other independent labels, both in France and abroad. Sixpack put out a split with Bay Area pop-punkers Samiam in 1996. Various bands connected to Meantime had opportunities to play with DC groups like the Capital City Dusters, who made a great impression on Sixpack's drummer Maxime Charbonnier. After his band dissolved in 2000, Charbonnier decided to complete the circle and head to DC on a holiday that would prove to be life-changing.

Crapoulet Records, a boutique label based in Marseille, prominently features an image from a Minor Threat show as a banner online. Olivier Firminhac, who moved south from Paris in 2008, founded the label using Dischord and the greater urban ethos of DC as templates. Though Crapoulet does release some music by artists from Marseille and Montpellier, their mission has grown to provide a platform for underfunded bands in countries like Argentina, Chile, Croatia, and Israel.

"For the label, I used to say that I'm running it like Dischord, because there is no contract, it is very simple," said Olivier. "The maxim is in Do-It-Yourself. It's not like Darbouka [Records], which [releases punk] records from Africa... [our] audience is somewhat less big, but they are more big fans of the music. It is the spirit I am trying to give."

His wife Claire Samant, who opened a tattoo parlor in 2013 on Boulevards Longchamps, named the shop after the Bad Brains' "Sailin' On." Though a punk fan since she was 14, she admits coming to hardcore relatively late, and that it took refreshing takes from the DC scene to attract her to it.

Claire Samant in front of her tattoo parlor in Marseille, named after the Bad Brains song. (Photo by the author)

"I did not get into hardcore until really late, because I thought it was stupid and [overly macho], you know, people dancing round and round [in a pit]. So when I started dating Olivier, he said he was into hardcore, and I was like 'please no!'" she laughed. "And then he made me listen to Black Flag and Minor Threat, and I thought, 'Oh, this is really cool.' So, I like the spirit, I guess. It is the more…sane? scene, I can figure, because what I know about New York hardcore, we call it 'hardcore de beouf.'"

"Redneck hardcore," Olivier clarified.

"Not redneck," countered Claire, "but just [in a husky, goofy voice] 'huh, we are real mean,' mosh pitting …What I think about [the DC aesthetic] is more common sense; they are more 'think wisely and do what is cool.'"

"And no [Hare] Krishna shit," laughed Olivier, firmly placing DC in contrast to their New York counterparts, many of whom got involved with the controversial Krishna consciousness movement.

In 2013, Crapoulet had the unique opportunity to release rare recordings by one of the seminal harDCore bands, and they decided to do so in a novel manner. Like many French hardcore fans, Olivier had been listening to Government Issue for much of his life. On a whim, he contacted John Stabb about a release he was planning to benefit a rabbit shelter in Marseille. Stabb actually did have some unreleased, scratchy demo recordings from 1982 and figured he may as well let them go to a good cause. The set included crudely recorded versions of tracks like "G.I.," "Rock 'n' Roll Bullshit," and "Anarchy Is Dead." The average track length was well under one minute. Olivier realized that anyone who would want a physical copy of these demos would really be in it for the collectors' item and good cause. Given how expensive it would be to press a special 7-inch record, how common cassette tapes had become, and how fans of the label could download the audio for free, he realized he could fit all of the tiny audio files across two 3.5-inch floppy disks. A boutique label who owed its existence to Dischord's influence now owed its most unique release to one of the foundational Dischord artists, a testament to the accessibility that has defined the DC scene for many international fans.

The animal-loving ethos of Crapoulet notwithstanding, this case study demonstrates the primary reason why indie labels circulate urban ethos and place identities so well: their owners want them to. Any record label based outside of a major entertainment market like Paris should not be expected to sustain itself by just selling products to kids in the neighborhood. In recounting the history of Dischord to the Library of Congress in 2013, Ian MacKaye rhetorically asked the audience what exactly they believed the larger music industry was selling. His answer was "plastic." Labels like Dischord, No Idea, Sarah, Postcard, Meantime, and Crapoulet are selling an image and a mentality embedded in DC, Gainesville, Bristol, Glasgow, St. Etienne, and Marseille, respectively. In other words, local record labels serve as shorthand for urban landscapes and become conduits for the circulation of local identities.

Bibliography

Adams, P. C. (2007). *Atlantic reverberations: French representations of an American presidential election*. London: Ashgate.

Andersen, M., & Jenkins, M. (2001). *Dance of days: Two decades of punk in the Nation's capital*. New York: Akashic Books.

Azerrad, M. (2001). *Our band could be your life: Scenes from the American indie underground, 1981–1991*. New York: Little, Brown.

Bell, T. (1998). Why Seattle? An examination of an alternative rock culture hearth. *Journal of Cultural Geography, 18*(1), 35–47.

Briggs, J. (2015). *Sounds French: Globalization, cultural communities, and pop music, 1958–1980*. Oxford: Oxford University Press.

De Tocqueville, A. (1978). *Democracy in America* (21st ed.). New York: Mentor.

Dunn, K. (2016). *Global punk: Resistance and rebellion in everyday life*. New York: Bloomsbury.

Foley, M. S. (2015). *Dead Kennedy's fresh fruit for rotting vegetables. 33 1/3* (Vol. 105). New York: Bloomsbury.

Frith, S. (1996). Music and identity. In S. Hall & P. D. Gay (Eds.), *Questions of cultural identity* (pp. 108–127). London: Sage.

Gillett, C. (1970). *The sound of the city: The rise of rock and roll*. New York: Outerbridge & Dienstfrey.

Goode, J. M. (2003). *Capital losses: A cultural history of Washington's destroyed buildings*. Washington, DC: Smithsonian Books.

Green, N. L. (2014). *The other Americans in Paris: Businessmen, countesses, wayward youth, 1880–1941*. Chicago: University of Chicago Press.

Kassabian, A. (1999). Popular. In B. Horner & T. Swiss (Eds.), *Key terms in popular music and culture* (pp. 113–123). Malden: Blackwell Publishing.

Kelly, M., Jones, T., & Forbes, J. (1995). Modernization and Avant-gardes. In J. Forbes & M. Kelly (Eds.), *French cultural studies* (pp. 140–182). Oxford: Oxford University Press.

Looseley, D. (2005). Fabricating Johnny: French popular music and national culture. *French Cultural Studies, 16*(2), 191–203.

Nelson, V. (2013). *An introduction to the geography of tourism.* Lanham: Rowman & Littlefield.

No Author Credited. (1977, April 9). Jazz Reissues become growing industry. *Billboard Magazine*, p. F-5.

Regis, H. A., & Walton, S. (2008). Producing the folk at the New Orleans Jazz and heritage festival. *Journal of American Folklore, 121*(482), 400–440.

Scriven, M., Hewitt, N., Kelly, M., & Atack, M. (1995). War and class wars (1914–1944). In J. Forbes & M. Kelly (Eds.), *French cultural studies* (pp. 54–96). Oxford: Oxford University Press.

Slickzine, 1985 (DC)

Smith, H. (2016a, May 9). John Stabb, punk rock headliner of D.C. music scene, dies at 54. *The Washington Post*. Retrieved from https://www.washingtonpost.com/local/if-these-walls-could-talk-theyd-probably-scream/2016/08/01/86bbed62-5751-11e6-831d-0324760ca856_story.html?utm_term=.26f1de1fea31

Smith, T. L. (2016b, December 9). Rock & Roll Hall of Fame 2017: Hard to ignore Bad Brains' influence. Cleveland.com. Retrieved from http://www.cleveland.com/entertainment/index.ssf/2016/12/rock_roll_hall_of_fame_2017_ha.html

Taylor, H. (2001). *Circling Dixie: Contemporary southern culture through a transatlantic lens.* New Brunswick: Rutgers University Press.

Tuan, Y. (1977). *Space and place: The perspective of experience* (10th ed.). Rochester: University of Minnesota Press.

Wallach, J. (2008). Living the punk lifestyle in Jakarta. *Ethnomusicology, 52*(1), 98–116.

Wetzel, R. (2012). *The globalization of music in history.* London: Routledge.

4

Washington Geography and the Birth of HarDCore, 1979–1983

Though a sizeable constellation of revered bands proudly carries the "DC" label, to comprehensively describe "the DC Sound" is impossible. The city's underground music scenes, though they often overlap due to the District's relatively small size, traverse several genres and draw inspiration from decades of niche musical cultures. The closest approximation to "a DC sound," as many of those interviewed for this book echoed, is an attitude closely affiliated with the social context of the music's production. The elements which are culturally bound with bands from DC are similar, ontologically, to "the Seattle sound." A few years after Seattle's time in the international spotlight and atop the *Billboard* charts had cooled, Tom Bell (1998) offered a cogent analysis of both geographers' and journalists' collective naïveté:

> Cultural geographers have tended to cling to the notion that there is such a thing as the personality of a region. This has been the case at least since the time of the French regional school that first popularized the notion; cultural geographers are not the only ones. Journalists would like to believe this as well and their frantic search for the meaning and even the existence of a Seattle "sound" is proof positive of that belief. If a Seattle "sound" did not really exist, they simply set about to manufacture one. To anyone who has listened to the music of the Seattle-based bands, the notion that they

© The Author(s) 2019
T. Sonnichsen, *Capitals of Punk*, https://doi.org/10.1007/978-981-13-5968-2_4

sound alike is almost ludicrous. They range from the folk-inspired Walkabouts to the quasi-heavy metal sound of Soundgarden, the heir apparent to the suburban Seattle based 'real' heavy metal bands Queensryche and Metal Church of an earlier musical decade. Journalists have been falling over themselves, however, trying to weave together disparate threads to create a sense of commonality about Seattle's music scene. The elements they have selected are the noise level of the music, its honesty, and the degree to which many of the groups were treated better elsewhere before being accorded their due in their hometown. (p. 37)

Seattle, while being the prototypical case study of the conflict between urban musical geography, the popular discourse, and marketing, was hardly the only application. Following Colin McLeay, Tony Mitchell had a similar assessment of Dunedin, and the public's fascination with its music generated "through a cultural geography of isolation, which produced a 'mythology of a group of musicians working in cold isolation, playing music purely for the pleasure of it' … [and] made Dunedin a metonym for Aotearoa/New Zealand music as a whole" (Mitchell 1996, 224). DC has accrued a comparable mythology, presenting an ideal of self-sufficient DIY network to which other scenes around the world could and seemingly should aspire. As previously mentioned, several artists borne out of the DC underground went onto mainstream success, so this is an obvious simplification and romanticization. That being said, DC's collective contribution to the canon of punk is one of few scenes so universally revered.

This was especially true by the time Seattle's sound hit it big in 1991 and cities like DC and Dunedin were suddenly viewed as stables for future pop stars, most of whom failed to reach their proscribed commercial potential. To tag one band prototypical of the DC scene would be both unfair and irresponsible. Fugazi were the most traveled and enduringly popular punk band from DC, but Minor Threat were perhaps the most emblematic by-product of what Craig Wedren refers to as the "big bang" of hardcore. Then again, Bad Brains have received the most recognition on an industry level, while many fans and critics prop up Rites of Spring, who played 16 shows, as the "moment" when DC punk reached its early artistic zenith.

The fast-and-loud style was not unique to DC, though DC put a brand on it, standardizing the subgenre's label as a declaration of dedication. The teenagers forming speedy bands in Bad Brains' wake called themselves "hardcore" punks as a reaction to the docile punk fashions and the poseurs who wore them around DC in the wake of punk's spark of popularity in the late 1970s. Kids like Ian MacKaye and Nathan Strejcek were serious about what the music meant, as well as the self-determination possible through it, rather than simply playing the part to seem edgy.

The trope of DC hardcore revolved around shouted anthemic lyrics, triple-time drumming (sometimes hovering as high as 200 bpm), and slashed power chords, wherein a guitarist's fretting hand could maintain a standard shape and position while frenetically sliding up and down the neck of the guitar (Easley 2015). Scholars like Charles Fairchild contextualized the harDCore aesthetic as a deconstruction of the traditional guitar-bass-drum rock band formula. The lead-ins, verses, bridges, and choruses were often present, just blitzed through with however little abandon their hyper-caffeinated authors had. The first four EPs that Dischord released in 1981, all on 45-RPM 7-inch vinyl, contained four songs per side. The label collected them all on one 33 1/3 RPM 12-inch record in 1984. The longest track on the whole collection was the Teen Idles' "Too Young to Rock," clocking in at an astronomical 2:04 (owing much of that to being a sloppy live recording). The longest studio-recorded track was Minor Threat's "Filler," at 1:31.

DC-style hardcore was not difficult to play, though it was a challenge to play well. Bands like Bad Brains and Minor Threat absolutely did. Even to the untrained ear and uninitiated punk listener, Minor Threat's sonic assault was sharply calculated and deliberately accentuated for maximum impact. Music theorist David Easley (2015) unpacked Minor Threat's intricate musical values to say how well the band's backing fits together with the messages in Ian MacKaye's lyrics, which were highly didactic and bent on self-determination.

While Minor Threat's music eventually slowed down into something that invited inevitable musical and personal differences that dissolved the band in 1982 (and again, for good, in 1983), their catalog of lightning-strike songs like "Straight Edge," "Out of Step," and "In My Eyes" has withstood the test of time. The group's legacy continues to swell globally decades after they played their final gig. The global impact of "Straight Edge," a 41-second song that MacKaye wrote about his sober outlook, is incalculable. According to some sources, the first traceable reference to "Straight Edge" in South America came on the cover of a 1982 compilation in Brazil called *Grito Suburbano*. Sao Paolo remains, perhaps even more so than DC or Paris, a center of contemporary Straight Edge culture (Reia 2015). According to Robert Voogt of Commitment Records, a Dutch label known for releasing music by Straight Edge bands, the movement took root in Europe through the late 1980s, rising within punk scenes in the Netherlands, the UK, Germany, Belgium, and Italy, with Scandinavia not far behind. The scene members closer to home were a bit less sold on it.

"So many American boys were fetishizing and romanticizing and beating their chests over their impression of the DC hardcore scene of 1981 that it turned into something a little bit religious in that way," reflected Craig Wedren, who moved to DC after Minor Threat had broken up. "[The whole thing] becomes dogmatic. When romanticizing or fetishizing turns into dogma, then you're starting to shut down the creative freedoms. And one could argue that there are some near-Fascistic movements like Futurism or metal that have produced some pretty awesome material."

Unsurprisingly, Straight Edge has landed well in cultures that already eschew alcohol consumption such as Muslims in Indonesia and Latter-day Saints in the Great Basin region of America (Dunn 2016). According to French straight edger Nicolas Gresser, however, Paris and France at large are relatively lacking in a cohesive community. He has often had to leave France in order to find hovels of that community in England or in the US. His last few US trips have been motivated by seeing Straight Edge hardcore bands like In My Eyes (named after the Minor Threat song) in Baltimore. He first discovered Minor Threat as a teenager via their *Discography* CD at a second-hand store in Dole, his hometown near the Swiss border.

"They looked pretty normal [like skaters and nerds], and drinking Coke, but their music was like the punkest music ever," reflects Gresser. "It was only a couple of years later when I [learned better English and] found out what Straight Edge was about. And it was funny because…I was like 'there, it all makes sense.'"

As the band has made a series of profound impacts on his life, so has Minor Threat's signature within the punk landscape. A Minor Threat T-shirt is likely to elicit compliments at a punk gig or record shop anywhere in the world, and an underground music scene in which Minor Threat's members and their friends grew has reached iconic status that has transcended the music itself.

"The sound now with all of the bands is totally different, you wouldn't believe it," Guy Picciotto told the German fanzine *Trust* in 1988, "there's no 'sound.' No DC sound at all now. It's really varied. The artistic levels are totally diverse. The things that hold the scene together are more like connections between friends, links like that, rather than some code of ethics or a sound, things like that. I think that is really healthy."

By the time that Shudder to Think left Dischord to sign with a major label, the band was based in New York City, where Wedren had been since he left DC to attend New York University in 1987. Their guitarist Chris Matthews, bassist Stuart Hill, and drummer Mike Russell remained in the DC area, though, and Wedren would take the train down almost weekly to practice and play shows. It was actually on the train between New York and DC where he wrote the majority of the band's lyrics. Shudder to Think remained solvent as a touring unit, and they were happy to carry the DC tag on the road. The city's urban ethos freed many from a musical orthodoxy. As the band told *Sidekick* zine in 1990:

Craig [Wedren]:	It helps [being from DC when we're on the road].
Chris [Matthews]:	It certainly gives us license to not be (plays generic 1-2-1-2 drum line on his knees) 'cause we're not into that that much. We'd much rather do what we do.
Craig [Wedren]:	And I think it makes people want to hear us. There are so many good bands that have come out of DC. They expect quality.

Live videos from that year of Shudder to Think playing early iterations of songs like "Red House" strikingly presaged the imminent grunge explosion. Though they were a DC band, their aesthetic matched that which would soon characterize on a grand scale so many of their contemporaries in Seattle: long, shaggy hair (at the time, at least), flannel, baggy pants, murky yet melodic guitar tones, and a stomping rhythm section. Though they sounded similar to some Seattle bands at the time, it was ultimately that "musical honesty" that tied them both together. Shudder to Think, typical of bands on the Dischord roster or within the label's orbit, were content to let their live show bleed through on the recordings, avoiding fancy studio effects or glossy overproduction.

This prompted Nick Crossley, drawing on network theory in his research on the late-1970s Sheffield, UK, post-punk scene to suggest that because many bands recorded in Cabaret Voltaire's studio, the whole scene was more vulnerable to the band's unifying influence. This bares a clear parallel with the DC scene, though over a wider time span. Don Zientara's Inner Ear Studios in Arlington has been the recording spot for dozens of classic DC records by groups of several classifications. For the label's first decade, many of the Dischord bands recorded there, and Ian MacKaye would often helm production. In addition to being the label founder, he aided the sound and production decisions of many bands that helped shape DC's sonic aesthetic. Multiple punk and indie musicians, including Smart Went Crazy's Chad Clark and Aloha's TJ Lipple, also moonlighted as recording engineers, eventually picking up that mantel full-time. J. Robbins, despite Jawbox being signed to Atlantic from 1993 through their breakup in 1996, also produced local recordings, eventually opening his own studio in Baltimore, the Magpie Cage.

This dynamic was common among smaller, self-contained scenes in France. In several cases, artists, engineers, and producers traded places between DC and Paris. After their initial visit to DC in 1987, two 20-year-old Parisian punks named Philippe Roizès and Arnaud Gabelli brought stories about Inner Ear back to their crew in Paris. The following spring, Roizès mailed MacKaye a letter indicating that the French punk band Flitox were planning to record at Inner Ear that August, and that Gabelli's band Cosmic Wurst were also interested in speaking with

Zientara. A few years later, Ted Niceley, who produced early albums by The Slickee Boys and Fugazi, found himself in high demand in France. In the early 1990s, he went over to produce a record by Fugazi fans Noir Désir, returning throughout the decade to work with other bands, culminating in one of his favorite projects by the DC-influenced Bordeaux two-piece Gâtechien.

"[They are] a two piece group consisting of Laureau Paradot on Bass and vocals and Flourian Beloud on drums and vocals," Niceley enthused to me in an email. "A ferocious yet melodic record. It was as if Fugazi and [The] Birthday Party had a jam session on this studio called Le Nef in Angouleme as this eerie fog rolled in and all of a sudden a lightning bolt struck and Gâtechien was the bastard child."

Of course, this raw-studio-sound ethos, which had been pertinent in DC, became coveted, standardized, and ultimately placeless by the early 1990s. Prior to the major-label-abetted explosion of Nirvana and the mainstreaming of "alternative" rock by proxy, neither Shudder to Think nor Mudhoney, Mother Love Bone, or even Nirvana were composing their songs with any concrete expectations. In fact, Dave Grohl admits that, upon leaving Scream to move to Seattle and join Nirvana, he considered Fugazi to be the paragon of a successful independent rock band. He hoped his new band could sell half as well as his hometown heroes; what happened by the end of the year was something few could have foreseen. Despite the meteoric rise of Nirvana and the scene they carried on their collective backs, most insiders posit that none of it could have happened from an artistic standpoint without the prior six years of fall-out from Revolution Summer in DC. The DC scene receives due credit for raising Grohl, who played a prominent role in making Seattle's packaged "sound" palatable (Crawford 2015). Many Seattle bands of that era also modeled their guitar work off of Black Flag's later records. Another DC punk, Henry Rollins may not have written Flag's music, but he was the face. This set of circumstances did not go unnoticed on an international scale, particularly by Parisian punk and indie veterans like bassist Roman Jaskowski.

"When grunge got big and the music arrived in Europe, it did nothing for me," said Roman. "I felt I had heard all that already, from DC punk.

That scene was preparing what was next – it was the youngest and freshest version of it."

Bad Brains and Hardcore Iconography

To understand the adversity which gave rise to hardcore punk in the US may have been difficult for French punks at the time, still relying heavily upon dominant British iconographies. Lyon punk veteran Maïe Perraud enthusiastically cites the culture cafes she encountered during her teenage years as pivotal in the proliferation of punk culture in France. The François Mitterrand administration, between founding the Fête de la Musique and sanctioning culture cafes nationwide, heavily contrasted with the Anglo-Saxon governments led by Ronald Reagan and Margaret Thatcher. Rock critics cited waves of sociopolitical conservatism under Reagan and Thatcher as the source of undeniable fomentation for a highly volatile, exciting musical underground. Shayna Maskell, in her treatise on Bad Brains, makes the following case:

> The meaning-making occurring at [the Bad Brains' well-documented 1982 performance at] CBGBs is a refutation of the staunch conservatism and strict moral turpitude professed and accepted as the social mores of the 1980s. While this dominant political view itself included violence—that of the State against other countries, and the State against minorities, both figuratively and literally—the violence of Bad Brains and their audience was a sort of reclamation of violence as representation. (2009, 414)

In 1982, three years after releasing "Pay to Cum," Bad Brains released their debut full-length album on cassette for ROIR. As the band was too fast and loud for mainstream radio play or MTV at the time, promotional efforts were stunted and the cassette became a holy grail for international fans and collectors. Over the years, the album's cover art, which featured a cartoon of lightning striking the US Capitol, bedecked with a color scheme befitting the band's newfound Rastafarianism, has become among punk's most referenced images.

Banned in Rouen 7-inch release by Tekken (Rouen), featuring lightning striking the Jean d'Arc Cathedral in homage to Bad Brains LP cover. From the collection of Fast Fab. (Photo by the author)

The Rouen band Tekken released a split 7-inch in 2006 entitled *Banned in Rouen* (a nod to Bad Brains' "Banned in DC"), lifting the ROIR artwork, replacing the Capitol dome with a similarly distorted image of the Jean d'Arc Cathedral. Thrashington, DC released a special edition of their 2007 LP *To Live and Die in B.M.O.* (Brest Metropole Océane) that included a poster of a cartoon showing lightning bolt striking Brest city hall. Incidentally, that album included an original song called "Banned in B.M.O." and a cover of Minor Threat's "Bottled Violence." I only included examples here relevant to the Franco-American punk story. The Bad Brains image has inspired parodies in the hundreds, possibly even thousands. In 2017, Asbestos Records, a punk label based in Connecticut, issued a T-shirt that aped the Bad Brains artwork to say "Banned in New Haven," replacing the Capitol dome with a pizza, New Haven's signature gastronomic export.

Thrashington, DC Insert Poster from Special Losers Deluxe Edition of "To Live and Die in B.M.O." (2007), with Lightning Striking the Mairie du Rouen à la the Bad Brains LP cover. From the collection of Fast Fab. (Photo by the author)

Bad Brains' musical eclecticism also diversified DC's early punk ethos beyond merely color lines. The early harDCore kids, following in the footsteps of a black band, forged a kinship with the city's Go-Go funk scene. Though segregation drove wedges between the two, they attempted to build bridges and occasionally booked shows together. Government Issue and Trouble Funk once played together in a historically genre-melding event. The results were mixed; many of Trouble Funk's older fans were less interested in harDCore, and despite the (relatively) large number of black faces at punk shows, the two scenes' fans did not mesh on a large scale. Also, though they both encouraged crowd participation, Go-Go's long-form jamming and call-and-response mechanics did not cohere with Government Issue's short, punchy songs. Despite the short-comings of the experiment, one predominantly white scene and one almost-exclusively black collectively put in more effort to bridge the widening racial gap of the height of DC's crack era than anybody at city hall. As Natalie Hopkinson wrote in *Go-Go Live: The Musical Life and Death of a Chocolate City* (2012):

> Go-go can feel a lot like a Pentecostal church service. Both run for extended hours, and you never know when they will end. Neither erects a huge barrier between who is performing and who is watching. Both are heavy in call and response. In the beginning there was [Chuck] Brown, then generations of Washingtonians came after him that continued to make go-go a place, subject, and verb. (p. 51)

Similarly, in the beginning there was Bad Brains, and then generations of Washingtonians came after them that continued to make hardcore a state of mind, a style of music, and an adjective. Hardcore, too, could feel like a religious ceremony with appropriate levels of devotion and tribalism. Shows would sometimes run for several hours and similarly stretch well into the night, provided the venue or kids involved had no curfew.

Meanwhile, in the musical mainstream in 1979, the racially motivated "death of disco" coincided with deflated mainstream interest in punk music. According to ushers at Comiskey Park in Chicago, few black fans attended Disco Demolition Night, and the white fans bent on destroying disco records were bringing in LPs by black artists who had never recorded a disco song. The music industry mirrored the public's tacit rejection of black music, simultaneously disinvesting in punk and disco. Coincidentally and ironically, the most successful bands which signed with majors during punk's first wave had reoriented their sound toward black music (Blondie with disco and The Clash with reggae and ska) or dissolved (The Sex Pistols and The Dead Boys), or were The Ramones. It bears a mention that 1979 was the year The Ramones recorded their glossy *End of the Century* with infamous producer Phil Specter. *End of the Century* has many acolytes, but it is hardly a representative piece of Ramones music.

As it happened, punk slid underground and gave birth to hardcore while disco slid underground and gave birth to house. The consequential mainstream ambivalence furthered an underground kinship between punk with disco, go-go, and reggae, a hybridization that ran as a strong urban-centric counternarrative to corporately controlled music media.

Ali Shirazinia and Sharam Tayebi, suburban Marylanders who would strike gold in 2000 as the house duo Deep Dish, cited Minor Threat and Bad Brains as early influences. Fittingly, many hardcore musicians in both DC and Paris transitioned into electronic music as the twentieth century went on. Manu Casana (the Sherwood Pogo vocalist whose 1984 visit to DC will be detailed shortly) transitioned into a successful career in House and Rave promotion. The cultural impact and meaning of House music circulation between DC and Paris would make an interesting counterpoint and companion piece for this story.

Despite both an understandable stereotype of the District as a button-down, vanilla atmosphere and its proud history as the quintessential African-American-dominated "Chocolate City," DC has long been a constantly shifting mélange of flavors that enhance the sweetness of one another. Though punk and hardcore are often subject to aggressive genre-fication, DC has problematized the dynamic that "the vast majority of musical production at any one time involves musicians working within relatively stable 'genre worlds'" (see Frith 1996), "within which ongoing creative practice is not so much about sudden bursts of innovation but the continual production of familiarity" (Negus 1999, 25). This played into the hands of the major labels, which historically prized and marketed formulaic music over innovative niche market subgenres.

Like many bands operating on the fringes of the music industry, Bad Brains had little reverence for genre. The quartet routinely performed dub reggae numbers in between their trademark fast-and-loud songs. Dr. Know incorporated heavy metal riffs early on, which moved the band into a more pop-metal style. While no constellation of bands can disprove Frith's idea of nebulous "genre worlds," DC did more than most punk scenes of note in that era to disrupt their stability. After Bad Brains set this precedent, many early harDCore bands like Scream, Void, and Red C, all of which included black members, followed suit. Through this and other types of transgressions against hegemonic ideas around a commonly misunderstood musical family, DC's entire history as a subversive "capital of punk" has consistently defined itself against the mainstream and repeatedly reinvented itself.

Gender and Urban Ethos of DC

Like many (though unfortunately not all) of their contemporaries in DC, bands like Fugazi made strides for social responsibility and political correctness in music, consistently defending marginalized groups. Though they were all white, middle-class men, their lyrics always punched upward to fight social injustices, especially within their own community. One such transgression was the aggressive and objectifying treatment of women in the predominantly male scene. As the feminist geographer Mona Domosh wrote: "When [women] move out of the house and on to the streets, our identities are constantly being monitored, judged, constituted, negotiated and represented" (1998, 280). Fugazi brought this to the forefront on their 1988 debut EP with the song "Suggestion":

> Why can't I walk down a street free of suggestion?
> Is my body the only trait in the eyes of men? (1989)

Fugazi were not the first male musicians to include feminist sentiments and experiences into their lyrics and turn the camera on their gender, but to the international community they forged an association of the DC scene with egalitarianism. Though few punk scenes were paragons of gender parity, DC concerned itself with providing a platform for marginalized voices that historically struggled to succeed on their own terms within the sphere of rock music (see Davies 2001). The band's efforts to convince their crowds to dial back violence to create a more inclusive space at their shows came to characterize (and caricaturize) the band.

While punk rock has never been an explicitly raced subgenre of popular music, relatively few progenitors have been people of color. Though on a worldwide scale, DC has produced many of the most notable musicians of color. Void guitarist Bubba Dupree, Beefeater guitarist Fred Smith, and Scream bassist Skeeter Thompson have all remained highly respected and influential throughout the hardcore underground for over three decades. Several black women also made profound impacts as the scene came of age. Red C bassist Toni Young and Fire Party drummer

Nicky Thomas are two prominent examples, the former also producing singles by local acts like Double O as well as managing bands before passing away in 1987 (Connolly et al. 2015, originally published in 1988).

Thomas' band, The Fire Party, was an all-female quartet active at a time when sexism in punk effectively birthed the Riot Grrrl movement in reaction. The most prominent Riot Grrrl bands (Bikini Kill, Bratmobile, and Sleater-Kinney, to name a few) originated in Olympia, but some figureheads including Kathleen Hanna spent the summer of 1991 in DC to build their eponymous zine and help sow the seeds of their radical feminist movement (Andersen and Jenkins 2001; Crawford 2015). Though the Fire Party had broken up by 1990, their members remained heavily involved in the community. DC had already been well established to international fans as a crucible of social justice in punk; it just happened that Fugazi were the most visible at the time, and "Suggestion" was the most audible. In 1994, Jawbox used a powerful platform afforded them by Atlantic Records to get their anti-objectification song "Savory" on MTV.

"One of the reasons, the driving force I'm in a band, is to try [to] show other women it can be done. You can do it. You can pick up an instrument and be in a band," said Jawbox bassist Kim Coletta in a 1990 interview with *Maximumrocknroll*. "For me it was kind of frightening at first because one of the reasons I didn't join a band for so long was I was quite frankly scared. Being female I thought 'if I don't play this bass fairly well people will just dismiss me as a female musician,' and I hate that … I'm a complete nut onstage now. It's really empowering because often, after I play, women come up to me, even if there aren't many women in the audience, the few there will come up and say, 'That was really great. It's inspired me. I've been trying to play guitar and I'm really going to try extra hard now.' That makes me so excited, almost more than anything else at a show."

Like the extensive roster of women who built that era of indie rock, Coletta was a musician first, but forced to operate within rock's masculine superstructure. One could ask the question "what is it like to be a woman in punk?" to 100 different people and receive 100 different answers. Women in punk scenes in France responded in kind.

"We were not all involved in the Riot Grrrl movement, [but] we were curious about them," said Maïe Perraud. "When I started to be in the punk scene, I never asked anybody to have my place to do this or that. I came here to do something. I never had any problems with men, maybe because I never mentioned that I was a woman, or wanted to do some-thing *as a woman*. I think that all the girls with me did the same. When we were in the pit, we didn't care.... I guess that if there are more punk rockers male than women it's because of history."

Few scenes in France, if any, are known for having large numbers of women at shows. This does not mean, however, that French women have made a less significant impact on punk culture in mainland Europe. Similar to their American counterparts, they play a prominent role in the circulation of zines both domestically and internationally. One zine distro, I Lost My Idealism, has been run by Gabrielle Casseville in her home in the suburbs west of Paris for years. "I learned how to play guitar when I was 9 until 16, the age when you're supposed to play in a band, then I stopped," Casseville said. "It's really different when you're a girl in the scene, because [you might be] too shy to play with boys, or... there is so much gossip, and when you're a girl it's horrible."

Regardless of how an individual may feel within their community and scene at large, punk increased this dialogue worldwide, and DC has been at the forefront of the charge. Though it was unfortunately not an area of wider focus in my research for this book, gender remains a key dynamic that permeates discussions about punk scenes, identity, and social justice in light of both DC and Paris. Both capitals are microcosmic of greater issues their respective countries face, especially regarding rights to bodily autonomy and justice for women and non-binary citizens. French women in their scenes, like their American counterparts, have found an avenue for self-actualization through punk.

"It's because of [French] culture," added Perraud. "Because we're in a very masculine culture, [but] it doesn't mean that women have [no] place in it. But you just have to *take* your place in it. I don't have to fight to have something, I just take it."

The Overlapping Geographies of DC Music Scenes

Interviewer: Do you think the way people fetishize DC of a certain era, do you think that's unfair?

Craig Wedren: I'm guilty of it; the older I get the more I romanticize '81, because I wasn't there [in DC yet]. I have no bad memories to ruin things… You can't appreciate it when you're in your early twenties, because it's your first time behind the wheel. But you can't think of 'this is our right, this is a meritocracy (laughs) everything's happening.' Then you look back and realize, holy shit, that was a real creative laboratory happening there.

As a by-product of the confluence of cultural spaces in and around the District of Columbia, harDCore did not develop in a vacuum. Nor did its creators intend for this to be the case. In fact, although many would not realize it at the time, DC's punk scene could never have happened were it not for an elaborate social, cultural, and internationally influenced history of their city. Though the nascent Paris punk scene was not exactly instrumental in the birth and early development of harDCore, DC's intermittent touches of Parisian influence appeared throughout the story of its birth and worldwide growth.

One unpredictably rich archive of documentation on DC's punk history was in the decidedly mainstream form of newspaper. The *Washington Post* and the *Washington Times* have paid attention, with modulated consistency, to the city's indigenous music scene despite an expected preoccupation with DC's governmental and civic life. Though punk would likely have grown in DC if the local press had never existed, the paper did immensely help. When punk was in the pipeline, the *Post* became an international sensation when reporters Bob Woodward and Carl Bernstein (as the party line goes) took down the administration and led Richard Nixon to resign. The paper's iconoclastic bent and radical undercurrent as a pillar of "the fourth estate" has maintained since then, even if physical circulation of the paper has declined and many of its counterparts around the country (e.g. the *Seattle Post Intelligencer*) have closed up physical shop.

When harDCore was young, the *Post* took an interest in the nascent scene around Georgetown. The old-money neighborhood was DC's first to gain federally observed historical status and was one of the first in the country to do so (Goode 2003), but that upper class was not the only facet of Georgetown life that the *Post* covered. In 1981, Cynthia Connolly, then a teenager living and checking out punk shows in Los Angeles, happened upon a *Post* article that mentioned the Georgetown Punks. Her recently divorced mother, accepting a job with the incoming Reagan administration, faced the prospect of moving her daughters to DC that year. Cynthia begged her mom to find a place in the neighborhood; they settled on a place on Hawthorne Street in Glover Park, less than a mile from Beecher Street. Cynthia brought her love of punk and her camera to the District, and the rest is history.

While the *Washington Post* has been cognizant of the local punk scene for four decades, the internet has spent the past two removing space restrictions from the content they can produce. Though citations of the local scene back then were rare, today stories and anecdotes on local punk history appear almost weekly on the *Post*'s website. For example, when contractors broke through walls at 2318 18th Street NW, they revealed graffiti left behind from 1979 and 1980, when the building housed the original Madam's Organ collective. In 2016, folk pop singer Ryan Adams posted a photo with Ian MacKaye on the Dischord House steps (a nod to Glen Friedman's iconic 1981 photo of Minor Threat) to his Instagram page, which the *Post Express* commuter paper reprinted. When Government Issue singer John Stabb (nee Schroeder) died of stomach cancer at 54, the *Post*'s music section wrote a touching tribute to him, detailing how he first got involved in the early Georgetown hardcore scene:

> The tony Washington neighborhood had become the center of a punk movement known as 'harDCore,' in which songs were rarely longer than two minutes, were screamed rather than sung, and frequently targeted corporations and commercialism. Teenagers with leather jackets, chains and boots walked the neighborhood's M Street drag in packs and—too young

to get into most clubs—made the Haagen-Dazs ice cream shop their home. (Smith 2016)

This counters a common sentiment often expounded by critical urban geographers who infer that newspapers "work to reinforce state-controlled templates that urban explorers are working so hard to reconstitute and reinterpret" (Garrett 2010, 1450). Newspapers still have the responsibility to document their city, nation, or world. Despite the questionable impacts wrought by corporate ownership, they serve, on a macrocosmic scale, the same ostensible function as boutique record labels or even fanzines. It is unlikely that any public memory or imaginary of DC will shift without some documentation thereof in the local press, especially considering how accessible local papers have made the punk scene to outsiders for a long time.

The more one reflects on DC's musical legacy, the harder it is to believe that DC is still characterized as a relatively a musical city by the outside world. Duke Ellington and Francis Scott Key both have their name on one bridge and one school each in the District. The two are the most historically prominent musical figures in DC history, but in manners that serve two conflicting narratives. Key was not even a musician per se. While Ellington wrote hundreds of songs that became jazz standards and almost single-handedly legitimized the form within American popular music, his name merely scratches the level of Nationalist clout as that of an amateur, slave-owning poet lawyer who wrote what became the national anthem. Of course, Key (1779–1843) and Ellington (1899–1974) lived during two dramatically different eras, and Ellington left his native DC behind at a young age to join the zeitgeist of the Harlem Renaissance. Key spent most of his life in Washington, practicing law and interacting with the federal government. Despite the preponderance of his name in nationalist discourse, Key's residence at 3518 M Street NW (1802–1948) became a casualty of Georgetown's postwar development. Though early tourism efforts converted the house into a museum in 1907, it eventually came down when the city completed the Whitehurst Freeway and connected it to the Key Bridge, completed two decades prior (Goode 2003).

Philippe Roizès (age 20) standing on the Key Bridge, Summer 1987. (Photo courtesy of Roizès)

Though both names attract a good modicum of interest from tourists, Ellington and Key still demonstrate why Washington, DC, does not *need* musical tourism. If there is one thing DC does not lack, it is tourists. In 2014, approximately 20 million tourists spent almost $7 billion in the city. Visitors to the District area engage in almost every type of known tourism, including political (e.g. a presidential inauguration), heritage (e.g. The White House, monuments), thano- (e.g. Civil War sites, Arlington National Cemetery, Vietnam Memorial), sports (e.g. college or professional basketball games), food (e.g. Ben's Chili Bowl, Old Ebbitt Grill), and educational (e.g. The Smithsonian Museums), often several within the course of one well-planned trip. Quotidian life for District residents has become inextricable from the urban tourist landscape, as "the distinction between tourism and other spheres of capitalist accumulation is fuzzy and imploding" (Corkery and Bailey 1994, 493). It follows that DC's size, density, and intersection of races and ethnicities across performance venues would mix that pot.

Perhaps the city's most prominent and politicized indigenous musical form, Go-Go funk music, is the greatest encapsulation of DC's ethnic site and situation. Though the DC punk scene was not monochromatic, it would be both inaccurate and irresponsible not to discuss Go-Go, a percussion-centered subgenre of funk indigenous to the district. Something as simple as music blaring from a car radio can alter the atmosphere of an entire city block (see Wissmann 2014), and recently local radio stations have ramped up efforts to keep Go-Go relevant in a gentrifying city. On a recent trip to DC, I had a related revelation while crossing the intersection at H and 9th Street NE. The Atlas District was full of activity that Friday night, but this block I was on was eerily quiet. Suddenly, a black sedan rolled by, blaring the familiar percussion-heavy sounds of Go-Go out of its windows. Within moments, the desolate block had transformed into a signpost of DC's great contribution to the funk music canon. As Natalie Hopkinson chronicled in *Go-Go Live: Musical Life and Death of a Chocolate City*, the District has been, for the past 40 years and throughout the punk era, a "laboratory for social innovation" (2012, 148):

> There is no better place to see [Habermas'] theory [of public spheres] in action than in Washington, D.C.: on the House and Senate floors, at the memorials for the Vietnam and the Second World Wars on the National Mall, in the Smithsonian Museums – and in the go-go clubs. These physical spaces are repositories of our national collective memory and the nexus of debate for the pressing issues of the day, whether they concern national governance, education, policy, slavery, poverty, or civil rights. The country's racially fraught history is the reason why the Chocolate City arose at this same intersection. Washington and D.C., go-go and 'mainstream' music public sphere appear separate and diametrically opposed. In reality, however, they overlap. Federal Washington and black D.C. are two sides of the proverbial coin. (Hopkinson 2012, 148)

As Go-Go has grown in direct opposition to the mainstream purview of DC, so has punk. Though the progenitors of punk came from more privileged or suburban backgrounds than the members of Trouble Funk or Experience Unlimited, they were all people, agents of circulation (mostly internal) acting as geographic mile-markers in a city fraught with contradictions.

Bibliography

Andersen, M., & Jenkins, M. (2001). *Dance of days: Two decades of punk in the Nation's capital.* New York: Akashic Books.

Bell, T. (1998). Why Seattle? An examination of an alternative rock culture hearth. *Journal of Cultural Geography, 18*(1), 35–47.

Connolly, C., Clague, L., & Cheslow, S. (1988). *Banned in DC: Photos and anecdotes from the DC punk underground* (7th ed., pp. 79–85). Washington, DC: Sundog Publications.

Corkery, C. K., & Bailey, A. J. (1994). Lobster is big in Boston: Postcards, place commodification, and tourism. *GeoJournal, 34*(4), 491–498.

Crawford, S. (Writer). (2015). *Salad days: A decade of Punk in Washington, DC (1980–1990).* New Rose Films.

Davies, H. (2001). All rock and roll is homosocial: The representation of women in the British rock music press. *Popular Music, 20*(03), 301–319.

Domosh, M. (1998). Geography and gender: Home, again? *Progress in Human Geography, 22*(2), 276–282.

Dunn, K. (2016). *Global punk: Resistance and rebellion in everyday life.* New York: Bloomsbury.

Easley, D. B. (2015). Riff schemes, form, and the genre of early American hardcore punk (1978–83). *Music Theory Online, 21*(1), 1–21.

Frith, S. (1996). Music and identity. In S. Hall & P. D. Gay (Eds.), *Questions of cultural identity* (pp. 108–127). London: Sage.

Fugazi. (1989). Suggestion. On *13 Songs* [Compact Disc]. Washington, DC: Dischord Records.

Garrett, B. L. (2010). Urban explorers: Quests for myth, mystery and meaning. *Geography Compass, 4*(10), 1448–1461.

Goode, J. M. (2003). *Capital losses: A cultural history of Washington's destroyed buildings.* Washington, DC: Smithsonian Books.

Hopkinson, N. (2012). *Go-go live: The musical life and death of a Chocolate City*. Durham: Duke University Press.

Maskell, S. (2009). Performing punk: Bad brains and the construction of identity. *Journal of Popular Music Studies, 21*(4), 411–426.

Mitchell, T. (1996). *Popular music and local identity: Rock, pop, and rap in Europe and Oceania*. Leicester: Leicester University Press.

Negus, K. (1999). *Music genres and corporate cultures*. London: Routledge.

Reia, J. (2015). I've got straight edge: Discussions on aging and gender in an underground musical scene. In P. Guerra & T. Moreira (Eds.), *Keep it simple, make it fast: An approach to underground music scenes* (Vol. 1, pp. 125–134). Porto: Universidade do Porto – Faculdade de Letras.

Smith, H. (2016, May 9). John Stabb, punk rock headliner of D.C. music scene, dies at 54. *The Washington Post*. Retrieved from https://www.washingtonpost.com/local/if-these-walls-could-talk-theyd-probably-scream/2016/08/01/86bbed62-5751-11e6-831d-0324760ca856_story.html?utm_term=.26f1de1fea31

Wissmann, T. (2014). *Geographies of urban sound*. London: Ashgate.

5

Hardcore Vient à Paris, 1983–1987

Owing much to its global orientation, romantic idealization, and commodification, Paris is regarded as much more musical than DC. The city's century of recognition as a hovel of jazz culture certainly helps, too, but I can't remember ever hearing Paris coming up in a casual conversation about punk. It's not that it doesn't have the necessary characteristics of a classic punk breeding ground: a history of working-class oppression and a high concentration of avant-garde thinkers and artists (see also: the *birthplace of the term 'avant-garde'*), and in an omnipresent characteristic, the city is *loud*.

Geographers have explored the phenomenology of urban sounds as building blocks of landscape, including more recently Torsten Wissmann (2014), who claims that "we do not listen actively to [urban sound, because] it is part of the urban environment and, therefore, taken for granted" (p. 22). Paris sat at the vanguard of many of the infrastructural transformations that made cities bigger, louder, and noisier. Railroads in the nineteenth century and automobiles in the twentieth century added harsh layers of mechanical noise to an already-crowded urban landscape. As Walter Benjamin wrote in *The Arcades Project*, "Paris is built over a system of caverns from which the din of Metro and railroad mounts to the surface, and in which every passing omnibus or truck sets up a

© The Author(s) 2019
T. Sonnichsen, *Capitals of Punk*, https://doi.org/10.1007/978-981-13-5968-2_5

prolonged echo" (1999, 85). I once heard British urbanist Matthew Gandy say that the "urban soundscape becomes the acoustic realm of late capital" (2016). This further echoes older sentiments from the prominent Marxist geographer David Harvey, who wrote extensively about capitalism's tendency to develop, destroy, and redevelop cities in its own image (1985, quoted in Berland 1992).

All of my interviews with informants in Paris reflected and worked around these capitalistic tensions. In two cases, my informants and I needed to relocate because our initial spot was too loud. It felt appropriate. It would be similarly appropriate if there existed more accounts of the Parisian hardcore scene, too, since Paris is a loud, chaotic city oftentimes posing major disruptions to capitalism. But, in one of the city's many contradictions, Paris created symbolic roadblocks to the point where hardcore had to operate out of the city's suburbs. The city made it hard for hardcore to exist, so telling the story is not without various challenges.

Qu'est-ce que le "Hardcore"?

Although this book may use the term "hardcore" liberally when labeling the Parisian music scene in question, it will focus on those bands and related actors whose sensibilities derived from American urban hardcore aesthetics. Bands that have been described as hardcore or hardcore punk have transcended multiple musical eras, and many of them share little similarity musically other than their ostensible genre worlds (Frith 1996; Negus 1999). Although the vast stylistic and aesthetic umbrella of punk music has enjoyed some academic engagement over the past four decades, the hardcore offshoot has seen relatively little.

The earliest cornerstone academic literature on punk, Dick Hebdige's *Subculture* (1979), staged it as a phenomenon of postwar youth culture in the UK and avoided pontificating extensively on the music directly. Despite the indispensability of *Subculture* for intellectual understandings of punk, reggae, and related subcultures, Hebdige relied heavily upon the most prominent imagery fostered by those genres' mainstream high water marks at the end of the 1970s. He necessarily excluded French punk from his analysis, though he did heavily cite French thinkers like Roland Barthes, suggesting that the punk ethos was a logical extension of the semiotician's philosophical output.

Reflecting on it in 2012, Hebdige never considered *Subculture* comprehensive or definitive, later lamenting "[doing] cultural studies proper a disservice by placing a gaudy, cartoonish wrapper on a serious activist and scholarly field of endeavor" (p. 401). He would also cite lucky timing for getting his manuscript accepted and the pressures of publication for its ostensible lack of oversight. Because the book came out in 1979, it missed the opportunity to look at the bevy of UK groups which profoundly expanded the crossover between punk and reggae by the end of the decade, particularly the second-wave ska groups like The Specials, The Beat, Madness, and the Selector—all of whom first released music that year.

Similar to disco's sublimation into underground dance music and rave, the fastest, loudest element of punk quickly slipped underground and reemerged in urban landscapes all over the English-speaking world. The term "hardcore punk" emerged in a tribal context that was inextricable from the cultural geography of American cities like Washington, DC. Divorced from its original sociogeographical setting, hardcore becomes more difficult to contextualize and define. British bands like Discharge and the Exploited began using the term to categorize a new breed of punk-metal crossover that grew to prominence in the UK around the same time. Many of the newer faster bands popping up throughout Europe in the mid-1980s were influenced equally by Dead Kennedys, Minor Threat, and Discharge—three distinct sounds coming from three disparate geographic settings.

Accurately tracing hardcore, especially by the time it unraveled in DC and gained traction overseas, is exceedingly difficult. As soon as hardcore became established as a subgenre in the US, it almost immediately began to splinter into other genre worlds like metal or became refracted into its more technical, confessional nephew, emo (née "emotional hardcore"). This is typical of noncommercial music, as the culture industry exercised little control over the music's transgressions.

"Mapping metal, especially its active 'underground,' is a messy task at best," wrote metalologist Deena Weinstein in 2011. "No laws or sharpshooting border guards keep bands playing within one style, nor are there any official music guardians or academic gatekeepers enforcing the standardized usage of terminology by critics, publicists, or fans. Moreover, styles are not watertight containers: they leak, bleed into others. Musicians borrow and steal, and styles constantly evolve and transform into new styles […] not even fans or critics know where to draw the lines" (p. 41).

Although occasionally given labels in other languages tantamount to a literal translation, "hardcore" has retained its original English terminology throughout dozens of languages in which it is played. Most French zines like *Positive Rage* and labels like Crapoulet referred to bands that mimicked the Minor Threat or Bad Brains sound as "punk hardcore" or "hXc" in French-language write-ups. Also noteworthy is, among French punk fans more so than American fans, "punk" and "hardcore" are distinct entities. Though hardcore undeniably branched off of punk, hardcore became more of a distinct subgenre to French fans, rather than simply a wing of the genre. The French underground had been saturated with Oi!, a quintessentially British punk offshoot that prized a working-class aesthetic that gradually became associated with the National Front and political hard-right (Worley 2014). Oi! was also generally simpler, slower, and prized gang vocals. The title was a reference to a cockney fighting incantation; if someone yelled it at you, you were probably about to wind up bloody. It was the battle cry of a generation unskilled in measured diplomacy. Violence was all they understood. Hardcore sped away from Oi! and, at least in France, typically eschewed the type of tribalism that tarnished their subculture. Due to the timing of hardcore's arrival in Paris, the subgenre quickly became a way to differentiate oneself from this creeping, regressive nationalism that had subsumed skinhead culture.

Hardcore Landscapes of 1980s Paris

The first skinhead in Paris was an Algerian teenager named Farid. He spent his childhood believing his father's plans to return their family to Algeria as soon as they had the means. They never did. Instead, they gradually put roots down in Colombes, a commune a few kilometers northwest of Paris, with no end in sight to their citizenship in postcolonial France. While Farid came of age, punk boiled over across the Channel from England. The sound crossed into France early with bands like Rouen's Olivensteins (named after a doctor known for helping addicts kick heroin), but it wasn't the first wave that captivated Farid and his friends Ammour, Pierrot, and Fan. After punk created a major-label feeding frenzy in England by 1977, the cockney-influenced Oi! subgenre was born, streamlined by groups like the Cockney Rejects, the Business, and, perhaps most important to Farid and his Colombes crew, Sham 69.

Oi! arrived at the perfect moment in Paris. Gérard Miltzine wrote to the Los Angeles–based zine *Flipside* in 1983 that at the end of the 1970s the punk movement in France became "a musical desert." French national television imported new wave culture from England on the show "Chorus," which showcased videos by bands like The Clash, Magazine, and Siouxsie and the Banshees. Los Angeles punk bands with international distribution like the Germs and the Avengers also eked their way through. However, to the diverse first wave of French skinheads, Oi! brought a new aesthetic that suited them. In Farid's words, "the uniform was clean, square, and scared people off." The first four French skins were not as hell-bent on destruction as recognition and self-identity.

Oi! music and the interwoven skinhead culture were largely apolitical until the early 1980s. Across the Channel, in England, the National Front rose to power and captivated the imagination of angry, disenfranchised youth; the nationalist and white power lyrics that Oi! bands like Skrewdriver adopted exacerbated their intolerance. In Paris, Le Front National grew under the leadership of Jean-Marie Le Pen, coalescing far-right sentiments against immigration and the supposed mongrelization of Western civilization, of which they viewed France as the apex. Multiple nationalist skinhead bands popped up around Paris, including the Tolbiac's Toads, Evil Skin, Brutal Combat, Bunker 84, and Legion 88. "88" was the commonly acknowledged shorthand for "Heil Hitler"; H is the eighth letter of the Roman alphabet. This became common because Nazi semiology like the swastika was illegal in postwar Germany. The Toads, named after their home neighborhood of Tolbiac in the 13th Arrondissement, rose out of the third generation of skinheads in Paris, the first to have overt nationalist and racist views.

The first two Oi! bands in Paris, L'Infanterie Sauvage and R.A.S., both broke up in 1984. R.A.S. decided to call it quits after too many of their shows ended in violence between nationalist skinheads and anti-Fascist punks. L'Infanterie Sauvage, however, disbanded because of related tensions within their own band. Their singer Géno (born Jean-Christophe Mam) moved too far right politically, engaging with the RAC (Rock Against Communism) movement and aligning with neo-Nazi groups. Remarkably, Géno was half-Cambodian, as were many young Parisian suburbanites in the wake of the diaspora caused by the Vietnam War and subsequent rise of the Khmer Rouge. Though these bands and many of their followers had fallen into nationalism as the by-product of "confusion" and "teenage politics" and would eventually turn against racism, their crews were responsible for most of Paris' exceedingly violent street gang activity.

The first wave of skinheads within the periphery of Paris appeared in 1978 around the Fontaine des Innocents in Les Halles in the 1st Arrondissement (Vecchione 2008; Beauchez 2014). The second wave emerged in 1980, centralized on the landmark fountain at Place de la République, less than two kilometers east of Les Halles. Still relatively unfazed by politics, this crew rallied around bands like Swingo Porkies. By the end of 1980, skinheads and divergent sectors of punks were making their geographic impact felt in downtown areas across Western Europe. In Paris, such rendezvous sites included Les Halles or in front of Beaubourg.

French skinheads were a diverse group in their early iterations, even including notable black and Jewish members. When nationalism and racism hijacked the skinhead movement, black gangs like the Del Vikings, Black Panthers, and Black Dragons emerged from the Parisian suburbs to fight back. All had associations with the punk scene. Extreme-right, nationalist, and racist elements would enter skinhead culture from various sectors of Paris and France at large. Even Farid, a North African who endured harassment and discrimination in an intensely diversifying Paris in his youth, developed his own take on French nationalism to rebel against his father and turn people's heads.

Today, Farid spends his days wandering around Menilmontant, meeting up with friends and answering questions about the old days from curious locals. Many are surprised that he, a former heroin addict and HIV survivor, is still alive, considering how badly the drug crippled that generation. Belgian National Television (RTBF), Planét Cable, and Cinétévé collaborated on the 1995 documentary *HAMSA, La Rage au Ventre*, following Farid and his friend Pierrot, then in their early 30s and still shooting up consistently. It's a challenging watch. Jérôme Beauchez (2014) wrote of that generation:

> The most radical become the most disillusioned upon the loss of standard nihilistic resistance which, certainly, results in a form of deadly hyper-consumption. Cannabis, beer, and wine are well-sold together with varied additives: amphetamines, hallucinogens or opiates, all ultimately intended to fabricate the pleasure and the strong sensations that the embrace of the banal and of the bad stifling fate of the everyday. (p. 194, translation my own)

Farid is one of few remaining signposts of the Oberkampf neighbor-hood's sordid past. One hundred years ago, Menilmontant and Belleville fostered elements of the city's premodern era, still full of cottages and rustic gardens. By the 1980s, the aesthetic had shifted to something much less bucolic. The vacant, blighted storefronts that once filled the street have given way to a French hybrid of gentrification and "embourgeoise-ment" over the past 20 years. The term "gentrification" became promi-nent in French urban reportage in the late 1980s, though the process had been active for over 20 years by that point. Belleville, which lies directly north of Menilmontant and runs over to Rue Pixérécourt on its eastern edge, resisted most city-coerced redevelopment until recently. Where the Belleville neighborhood was, as of a decade ago, "made of social housing and rundown buildings, where working class immigrants from Northern Africa and China lived" (Clerval 2008), the dominoes of development have fallen on this area recently. Granted, the city has had designs on it for years. According to geographer Neil Smith (1996), Belleville in the mid-1990s was a "solidly working-class neighborhood in the northeast-ern outskirts of Paris…a major center of Arab, African, Chinese and East European immigration and a traditional stronghold of proletarian oppo-sition" (p. 180). The city already had "amelioration schemes" and designs on the neighborhood at work by that point. In more recent years, establishments like the Belleville Brûlerie (cofounded by DC expat David Flynn) on Rue Pradier and Café Charbon on Rue Oberkampf have been de facto flagships of the neighborhood's transformation into an upper-middle-class hovel. Thirty years ago, it would have been unrecognizable.

Philippe Roizès told me as much as we walked down Rue Oberkampf, toward the Menilmontant Metro station in the 11th Arondissement. Thirty years ago, suburban punks like Roizès were unable to get from the Metro station two blocks away to hardcore matinees at Le Cithéa without being confronted by roving skinheads. Roizès recalls confrontations of his own as well as altercations he witnessed or heard of second-hand that turned violent or fatal. Although he knew the earliest skinheads in Colombes, Phillippe was too young to fall in with their crew. As skinheads became more rampant and violent in Paris in the early 1980s, their presence irrevocably influenced how punks navigated the urban landscape. On a smaller scale, Roizès and his school friends would steer clear of New Rose whenever there were skinheads hanging out in front of the boutique. They would duck away and return later, hoping the thugs had gotten bored and moved along. As the 1980s progressed, right-wing street gangs grew and this tense dynamic expanded citywide. Young punks, especially those known to affiliate with anti-Fascist gangs like the Red Warriors and Ducky Boys, needed to strategically plan their Metro trips throughout the city. Les Halles, Luxembourg, and the Clignancourt Flea Market were among the most dangerous stations to cross at the wrong times (Vecchione 2008).

The rampant street violence, compounded with the unfavorable timing of the shows, made the fledgling hardcore scene increasingly fragile. Because bands like Kromozom 4 and Heimat-Los were based in the suburbs, many of their friends had difficulty making it into Paris to see them. Because most hardcore shows in after-hours pubs could not even begin until 21:00 hours, most of the suburban punks had to leave by midnight in order to catch the last trains home or they would risk having to wander the city until 5:30 a.m.—a hazardous proposition, especially on a weeknight.

Occasionally, the hardcore scene found havens in venues willing to host matinee shows, such as La Cithéa, a dive on Rue Oberkampf and Villa Gaudelet. Because many of the bands and fans were minors (the aptly named Minor Threat being the quintessential example), venues that were normally 21 and over allowed clusters of bands to play all-afternoon shows. CBGB's in New York was one noteworthy example, fostering bands like Gorilla Biscuits which were too young to drink (despite how, like their Straight Edge counterparts Minor Threat, they had no desire). In Paris, although the drinking age was lower and less militaristically guarded than in DC or New York, few bars were eager to let the notoriously rowdy crowds take over for the night. The earlier start time enabled the largely suburban fan base of the splinter scene to actually get to and from the city at a reasonable hour.

Still, suburban punks like Roizès were unable to get from the Parmentier Metro station two blocks away to Le Cithéa without being confronted by skinheads. Although many of the altercations in question happened over three decades ago and the neighborhood had changed dramatically, veterans of hardcore shows in eastern Paris at the time seemingly all have these stories. Once, Heimat-Los singer Norbert Mension was on his way to La Cithéa for one of these matinees. It was uncertain whether Heimat-Los were on the bill; most of the hardcore bands shared members, fans, and friends, and therefore most of those associated with the tiny scene would attend most any show that happened. The way that Roizès relates the story, Mension was walking toward the corner of Rue Oberkampf when he was accosted by a group of skinheads. After a series of harsh words and threats, fists starting flying, and Mension never made it to the show. Greatly outnumbered, Mension was beaten until he was bleeding all over himself. He retreated to the Metro to go to his parents' house and recover.

Another night months later, a couple of the same skinheads were hanging out on that same block. Accounts of this event (like most altercations of this gravity and this long ago) vary, but the story goes that they waged war on another punk heading toward the Cithéa. This punk, known for his involvement in anti-Fascist action, was used to getting jumped by skinheads, but on this night, he'd had enough. He went back to his cheap, hot, top-flight apartment nearby and returned to the scene with a gun. He shot two of the skinheads; the younger one, 16 years old, died. The other one, a few years older, remains in a wheelchair today.

The shooter was apprehended and sentenced to seven years in prison. His sentence was extended at least three times—once for attempting to start a fire, once for punching a guard, and once for trying to escape. He left jail after 22 years, stricken with HIV from sharing dirty needles in a notoriously poorly maintained French prison. Roizès claims he met the man in question shortly after his release. Apparently, he is still living in a squat and engaged with anarchist activities in the Paris Underground.

"It's funny, when I met some guys from the New York hardcore scene – Agnostic Front, Madball, and such – they were saying 'New York is really tough!'" Roizès added, mimicking a tough-guy posture, then relaxing into a knowing smile and laughing slightly. "Yeah, okay. We have our stories also."

At this point in our conversation, Philippe stopped to point out a shop on a nearby side street. "The owner of that shop was very active in the Communist party in the 1970s and 1980s. He was very pro-China, pro-Mao. Still is." We continued walking. "But don't bother going in there and trying to ask about it. He'll pretend he doesn't know what you're talking about."

Perhaps the greatest obstacle to chronicling these stories, considering how the Parisian hardcore scene simmered through the mid-1980s without reaching the boil their counterparts in DC had, is the difficulty or outright impossibility of verifying most of these stories. Oral history, while providing a crucial platform for those on the margins to tell their stories, must always contend with the selectivity of memory. Additionally, when it comes to contentious epochs, one's political slant will always paint their account, even altering their memory so even the most impartial recall can be impacted.

These accounts must be taken as testimony rather than historically proven fact. While verifying truths can be difficult, oral histories around early punk scenes are especially challenging and often confrontational. Although these sordid accounts may vary over time, they recall a moment in Belleville's recent history that profoundly contradicts public perception and interpretation of Paris' landscape. Although their details may not be perfectly recalled or completely verifiable, they remain some of the only tangible, experiential evidence that the neighborhood was so dangerous and so blighted 30 years ago. La Cithéa, at 114 Rue Oberkampf, and a Middle Eastern grille shop next door seemed like the only places open on that entire block. Today, that block is so upscale that many of the punks and skins who once wandered up and down that street would hardly believe it was the same place. As of July 2015, the building that once housed La Cithéa had become the classy club Le 114, which had closed down as of this writing. The Grille shop next door, however, has blended in a trendy sushi spot, adjacent to a vapor cigarette outlet. Even those who once terrorized that block have become gentrifiers. According to Roizès, all of the surviving Nazi skinheads grew up and drifted away from the far-right youth culture, coming to reject racism and even befriending veterans of the hardcore scene they brutalized back then.

Even more tragically, much of that generation of skinheads is dead and never had the chance to share their stories. According to Roizès, Farid contracted HIV during his time as a junkie, but is "somehow still alive." About three of the skins who patrolled Oberkampf back then are still around, and all the others that Philippe can remember are dead, mostly from heroin overdoses or diseases contracted from drug abuse. He remembers hearing that one was killed by his dealer for not paying.

Some accounts of the battles that ensued between nationalists and anti-Fascists, including Mark-Aurele Vecchione's eye-opening 2008 documentary "ANTIFA: Chasseurs de Skins" (*ANTIFA: Skinhead Hunters*), omit this chapter from the story. Many of the affected and dejected include people interviewed for the documentary, including Farid.

One of the great individual tragedies of that generation, however, was the premature death of Géno, who never had the opportunity to outgrow and renounce his submersion into Nazism. His friends with whom he'd grown up in Colombes, including Roizès, try to remember him as the playful kid from when he started L'Infanterie Sauvage, rather than as the confused young man who started hanging out with neo-Nazis in 1984. On June 17, 1986, having severed contact with all of his old punk friends, Géno drowned in the Loire River at age 21. Despite the troubling politics that broke them apart, L'Infanterie Sauvage are still considered Oi! trailblazers and a crucial building block in French hardcore.

Much about the Paris hardcore scene, and for that matter the greater French scene, doomed it to obscurity. Few French (or even French-speaking) punk bands have made a major impact in the Anglophone world, commercial or otherwise. The ones heard outside of Europe in the mid-1980s were most often relegated to single tracks on mix tapes or ROIR compilations. However, it would be unfair for the musicians and scene members to shoulder much of the blame. Many cultural factors stunted the potential for a flourishing hardcore punk scene in Paris. The elements that would coalesce into the local hardcore scene in 1984, after the DC hardcore scene had fizzled alongside Minor Threat's 1983 breakup, had been largely relegated to the suburbs—hardly nodal points for regional and international circulation.

Depending on who one asks today, the first Parisian hardcore band was either Heimat-Los or Kromozom 4. The former (whose name roughly translates to "without homeland") was the first to record, but the latter had a history that reflected the insular genealogy of hardcore itself. In early 1984, the first two Oi! Bands in the Île-de-France region, L'Infanterie Sauvage and R.A.S., broke up. L'Infanterie Sauvage bassist Arnaud ("Arno") and drummer Felius (Félus) decided to part ways with Géno. Meanwhile, R.A.S., a firmly anti-nationalist band from Colombes, decided to quit after failing to divert neo-Nazi violence from their shows. Guitarist Taki and singer Gaz, who had met Arno and Félus the previous year at a punk festival in Essonne, decided to combine their forces and speed up their sound. Kromozom 4 was born, and some would argue, so was the Parisian hardcore punk scene.

Over the next few years, Kromozom 4, Heimat-Los, and other shorter-lived hardcore groups tried as often as they could to play shows in and around Paris. According to Roizès, who eventually joined Kromozom 4 as a second vocalist, most of the Parisian punks were still hung up on UK's first wave and hated hardcore, which made it even more fun for him and his friends to provoke the older generation. However, issues inherent in skinhead culture and violence compounded and created new sets of obstacles for the emerging scene, as did the restrictions that Paris' urban landscape and civic dynamics placed upon them as budding artists and purveyors of the DIY ethos they had inherited from their American counterparts.

The schism between Oi! and hardcore is a prototypical example of how problematic differentiation can be in writing about underground music scenes. The former originated in the UK, which was the wellspring of much of the punk culture that spread into France, being much closer both geographically and culturally to Paris than the US (Briggs 2015). Some faster recordings by R.A.S. might be describable as hardcore had they been recorded or released under that pretense. However, Oi! has never shaken its English working-class habitus. American hardcore had a similarly tribal yet more middle-class aesthetic, which translated awkwardly into French both lyrically and contextually.

It is impossible to say whether Kromozom 4 were the first band in Paris or even France to sound the way they did, but they were unquestionably among the first bands to emulate the fast-and-loud style that had trickled across the Atlantic from DC and San Francisco. Kromozom 4 had taken a cue from Heimat-Los when deciding to shed Oi! for a speedier style. Norbert Mension was one of the many Parisian punks whose fascination with harDCore came gradually. When Heimat-Los first began playing in the fall of 1983, they sought to emulate the thunderous and fast-paced British band Discharge. However, Mension's collaboration with guitarist Francois L'Homer brought a deeper understanding and appreciation for the sensibilities of DC hardcore.

"What was new to me in most of these DC bands was the fact that their lyrics were more in connection with real life," said Mension. "I mean most of the hardcore bands wrote about refusing to conform and government policies. And sometimes it sounded like caricatures. DC bands wrote about true and deep human feelings. There was something more smart, introspective and romantic about it in a way."

Like many of his suburban contemporaries, Mension never had an advanced English education, and thus his understanding of the lyrics by Minor Threat, Bad Brains, and Beefeater also came slowly. Back then, these records were especially difficult to find, prohibitively expensive to import, and there was not a high enough demand for these obscure bands for the bootleg market to respond. The cassette (or "K7/"kay-sette") tape trade dominated the spread of the style, which was indicative of an era when consumer electronics were rapidly diversifying and reacting to consumer demands for portability. Indeed, as studio time was too expensive, cassettes were now making it attainable for punk bands to record and distribute themselves independent from industrial production. Although some French TV shows like *Musique California* had broadcast American bands like the Dead Kennedys nationwide, MTV would not arrive in Europe until 1987. Paris' first wave of hardcore had largely passed by the point Dischord had brokered a deal to get proper distribution in mainland Europe.

Unfortunate yet understandable timing aside, the history of Parisian hardcore, even among the handful of bands included, has been poorly documented. Most of the popular press literature on French punk focuses on the first-wave bands like Les Dogs, synth-punk innovators like Métal Urbain and new wave crossovers like Téléphone. Christian Eudeline (2002), who wrote *Nos Années Punk: 1972–1978* (Our Punk Years), drew criticism from underground denizens for his lack of acknowledgment (or, to some, complete ignorance) of smaller DIY scenes. However, because the French hardcore community has always been relatively smaller and more close-knit than its expansive American counterpart, the interpersonal path and circulation of musical ideas is possible to trace, however vaguely. This generates reliance upon oral histories, which are not only inherently fallible but also often contradictory (Angrosino 2008; Turrini 2013).

Parisian Geography as an Obstacle for Punk

The hardcore underground of the mid-1980s represents a powerful counternarrative to widely accepted imaginaries of the City of Light. Although sites like the Eiffel Tower and Notre Dame have been among the world's most heavily trafficked tourist destinations for almost a century, a vast majority of the actual landscape lies outside the touristic purview. Similar to that of pre-gentrification American cities, a handful of mangy venues available to hardcore bands sat in appropriately blighted, marginalized corners of Paris. In 1984, most established music venues in the city were still hostile toward punk and barely even acknowledged hardcore. In both cases, the shows would attract a rowdy element who would get drunk, start fights, and sour the venue toward booking this style of music again.

Dynamics like these were hardly unique to Paris, but for the punk musicians and fans spread around Île-de-France, it compounded their frustration. Many of the people who played in or went to see those bands did not even live in the city proper, but instead around a handful of radical working-class suburbs. Mension grew up in such a housing project run by Communist mayors consistently from 1946. Some other hardcore kids, like Philippe Roizès, were fortunate to attend school in the city alongside their wealthier counterparts. Others were not as lucky; many members of that scene grew up relegated to the suburbs and did not have as expansive English education available. The suburban punks, who wanted to dig into new sounds coming through on mix tapes (imported 7-inch records were prohibitively expensive), therefore related to the aggression of the music rather than any didactic messages in the lyrics.

Roizès, whose family had moved to Colombes in 1974, began attending intermediate school in Paris in 1979. Although he and his friends would frequent the nearby punk boutique, New Rose, after school, all he remembers being able to afford were assorted badges and (on rare occasion when he had saved enough) domestic 7-inch records. American music, especially punk, was difficult to obtain. In 1980, his cousin two years his senior made him a mix tape that included a few Dead Kennedys songs, which was where his love of punk, specifically hardcore, began. Though he and his friends were not conscious of it at the time, their newfound fascination with the American version of what they had been fed from the UK continued a time-honored tradition in French popular culture (Briggs 2015). To European postwar youth, American culture represented a fun counterpoint and veritable antidote to the British standardization of culture and class systems (Gillett 1970).

This is not to say that this generation of "les punks" lost their interest in Britain; after all, it was much closer and therefore more accessible. Much of the underground punk culture that was stewing in and circulating through London, both homegrown and imported from the US, was handpicked by French punks and brought back home personally. Hugo Maimone from Lyon, then 16 years old and visiting London on holiday in mid-1981, found a copy of the Teen Idles' *Minor Disturbance* EP in a cut-out bin for £1. This was likely one of the first DC hardcore records to wind up in France. The impact of early harDCore releases like that was glacial, especially in France. Fortunately for kids like Maimone, Mension, and Roizès, the French government was about to go through a radical shift that for a brief period made their cities feel like powder kegs of creativity and accessible platforms.

According to British Urban Geographer Matthew Gandy, a discharging motorcycle in the center of Paris could conceivably wake up 200,000 residents early enough in the morning. By the mid-1970s, Parisians began to realize the transformative and subversive power of radio, and due to the population density of their city, they could hit a relatively large swath of population with relatively few transmission resources. Peripheral stations like Europe 1 and Radio Luxembourg were among the first to report any of the 1968 rioting, as French State radio played down these events. The earliest pirate radio stations in France were run by environmental groups and were called Green Radios. In 1977, Bruce LaLonde pulled out a radio on national television and played one of these broadcasts from the 7th Arrondissement. Similar to how brown-outs in the Bronx that summer were a flashpoint for hip-hop culture, so was this big break for underground radio in Paris. Jacques Chirac, who was an avowed Gaullist at the time, won the city's first Mayoral election that year. The new administration greatly suppressed free radios, but that would not last long.

In 1981, the socialist François Mitterrand won the General Election, quickly shifting the Federal government to the left socially and economically. As part of the Mitterrand government's national arts programs, radio waves were largely deregulated. The newfound openness of the radio waves in the incredibly dense urban landscape had a profound and sensational impact on French youth. Although radio had played a vital role in French social and political life since it had arrived in the 1920s, the medium continually found manners through which to redefine and transform people's perceptions of the schism between private and public space. The early spread of hardcore throughout metropolitan Paris in 1984 depended largely upon one's social and physical mobility, but radio's sudden deregulation enabled radical programs to find their way into the homes of curious Parisian teenagers.

One of these curious kids was a 14-year-old bass player named Roman Jaskowski, whose fanzine *Kakofony* provided some early documentation of the circulation of American hardcore into France. Jaskowski started playing music with friends, quickly forming his first band, Razzle Dazzle. They routinely listened to the anarchist, free-format radio show *Radio Mouvence*, one of few outlets offering a glimpse into regional underground music scenes for casual listeners. The show had been among the first to play Oi! from around Europe as well as early recordings by Bérurier Noir. In 1985, Razzle Dazzle played their first show, live on-air on Radio Mouvence with Krüll and the long-running Montereau punk quintet Les Rats.

The incursion and growth of the American hardcore underground in Paris did not begin until its original mould had shattered. Like its stateside counterpart, Parisian hardcore also splintered into various factions within a few years of its establishment, its founding members moving onward or shifting sideways by the later 1980s. Coincidentally, as American post-hardcore bands like Fugazi began to tour more frequently in Europe, more barriers to circulation fell for French scenes.

Hardcore's first wave in Paris in the mid-1980s, despite its brief lifespan, illustrates many facets of that relationship between the urban landscape and the underground scene. For all of Paris' contemporary strengths as a worldly city, a melting pot, and the heartbeat of Western culture, the reality for underground and unconventional music did not fit into the greater marketed image. The city's class stratification and physical layout both generated barriers to the circulation of any viable hardcore scene, particularly 30 years ago, prior to much of the gentrification that has altered the landscape.

These lessons could be applied to other underground scenes in other cities. The local conditions that generate particular scenes and styles or local interpretations of other scenes and styles are all necessary to consider when seeking to understand how and why the music may define the place. Musicologists and social scientists with a musical focus cannot disregard these relationships, no matter how anti-mainstream or under-documented the scene in question may be. Perhaps French punk fans, despite coming from what many of my informants claimed was "not a rock 'n' roll country," can take solace in this. In fact, perhaps less heralded hardcore scenes like that of Paris could be the key to a deeper understanding. While DC hardcore has been expansively reissued and documented, and while foundational UK hardcore bands continue touring worldwide in varying iterations, the small but spirited confluence of Parisian hardcore bands presented an unfettered and honest interpretation of the style, less mitigated by outside cultural forces or the rose-colored glasses of history.

Bibliography

Angrosino, M. (2008). *Exploring oral history: A window on the past*. Long Grove: Waveland Press.

Beauchez, J. (2014). La rue comme héroïne: expériences punk et skinhead en France. *Anthropologica, 56*(1), 193–204.

Benjamin, W. (1999). *The Arcades project* (H. Eiland & K. McLaughlin, Trans.). Cambridge, MA: Harvard University Press.

Berland, J. (1992). Angels dancing: Cultural technologies and the production of space. In L. Grossberg & N. Cary (Eds.), *Cultural studies* (pp. 38–51). New York: Routledge.

Briggs, J. (2015). *Sounds French: Globalization, cultural communities, and pop music, 1958–1980*. Oxford: Oxford University Press.

Clerval, A. (2008). *La Gentrification À Paris Intra-Muros: Dynamiques Spatiales, Rapports Sociaux Et Politiques Publiques*. Paris: Université Panthéon-Sorbonne-Paris I.

Frith, S. (1996). Music and identity. In S. Hall & P. D. Gay (Eds.), *Questions of cultural identity* (pp. 108–127). London: Sage.

Gandy, M. (2016, April 13). *Cultural geographies annual lecture: Urban Atmospheres*. Paper presented at the Association of American Geographers Annual Meeting, San Francisco.

Gillett, C. (1970). *The sound of the city: The rise of rock and roll*. New York: Outerbridge & Dienstfrey.

Hebdige, D. (1979). *Subculture: The meaning of style*. New York: Routledge.

Negus, K. (1999). *Music genres and corporate cultures*. London: Routledge.

Smith, N. (1996). *The new urban frontier: Gentrification and the revanchist city*. London: Routledge.

Turrini, J. M. (2013). "Well I don't care about history": Oral history and the making of collective memory in punk rock. *Notes, 70*(1), 59–77.

Vecchione, M. A. (Director). (2008). *ANTIFA: Chasseurs de skins*. Documentary. S. Brucher & M. A. Vecchione (Producers). Paris.

Wissmann, T. (2014). *Geographies of urban sound*. London: Ashgate.

Worley, M. (2014). 'Hey little rich boy, take a good look at me': Punk, class and British Oi!. *Punk & Post Punk, 3*(1), 5–20.

6

This Is Not a Fugazi Book: HarDCore Comes of Age

For years, the words SURRENDER DOROTHY in thick black spray paint adorned a CSX Railroad Bridge that spanned the Washington Beltway somewhere between Connecticut and Georgia Avenue, where the District's northern tip blends into Montgomery County. The graffiti was an inside joke about the Oz-like Mormon Temple protruding from the trees in the distance in Kensington. It provided a momentary break from the standard bumper-to-bumper drudgery for commuters until a few years ago when local authorities painted it over.

Toward the end of 2014, a new word appeared, painted in wide black strokes on that same spot: FUGAZI. Nobody knows who did it, but the symbolism could not be clearer. More than a decade after their final show, Fugazi still carries an equally iconic and iconoclastic status, representing a powerful transgression against the domineering narrative of Washington, DC. While their hometown was a geographic lynchpin of the global capitalist superstructure, Fugazi never once made personal profit from playing a hometown show. Notoriously, they kept all of their shows all-ages and $5 at the door. This bucked not only DC's capitalist reputation, but also the city's overbearing bar culture. Though Fugazi were not a Straight Edge band (despite having a member who wrote the song in 1981), they were not content with aiding and abetting the alcohol industry by

© The Author(s) 2019
T. Sonnichsen, *Capitals of Punk*, https://doi.org/10.1007/978-981-13-5968-2_6

restricting their shows to the 21-and-over crowd. They toured internationally for almost 15 years and sold more records than any other DC-based band, even making a few cameos on the *Billboard* 200. Still, many motorists had no idea what that graffiti meant. Some motorists may have even flashed back to the Vietnam era, where the term "fugazi" ("fucked up, got ambushed, zippered in") first appeared, but punk fans knew its broader meaning and significance in terms of place and music.

Fugazi as DC Geography

Ian MacKaye has cited his family's archival tradition, as well as a general dedication to self-representation and cataloging prevalent in the punk scene. In a 2013 address to the Library of Congress about "digital cultural heritage," he spoke about his grandmother Dorothy MacKaye, who wrote a marriage advice column for *Ladies Home Journal* under the name Dorothy Disney. She would tape-record conversations with troubled couples, which Ian would discover and rummage through as a teenager. Though the MacKaye family had lived in DC for more than four generations, his mother's family had come from Georgia during the reconstruction, bringing a thick storytelling tradition with them. Both sides of Ian's heritage played no small role in his aim to document the musical ongoing of the District via the label he and his friends began in 1980.

Once Fugazi formed and started playing more seriously in 1987, the band's collective attitude about international touring came from coming of age in a paltry live music town.

"We were really pushing to play in Italy and also France," MacKaye said. "We really pushed to do those gigs. There was a real resistance from the German [promoters] and the Dutch [bookers], saying, 'There's nothing there,' and we were like, 'We wanna fuckin' go.' I think it was because we were from Washington, DC, and early on, bands would come to America and just play New York and LA. And so we have a sensitivity about being the place where people were like, 'Ehhh, nobody really plays there,' and we were like, 'Yeah, and that's why we're gonna play there.'"

The band's collective origin in one of North America's most tourist-heavy cities also informed their attitude and approach offstage when traveling internationally. MacKaye, Guy Picciotto, Joe Lally, and Brendan Canty were consciously devoted to exchange and coexistence with their host cultures. They were not interested in being an "American band"; they just wanted their hosts to see them as a band. They were all equally concerned with breaking down many Europeans' preconceived notions about the ways that American bands behaved in Europe, even on the indie circuit.

"There's an ongoing joke that the American embassy is really just McDonald's, because that's where Americans would go as soon as they'd get into another country," said MacKaye. "First of all, we never ate at McDonald's because we were all vegetarian, and second, we would never go to a fuckin' fast food restaurant. We'd want to eat the food of whatever's going on... Maybe it was the French sensibility; I think it was just being a world citizen sensibility."

Considering how few American punk bands, especially independent ones, were touring France at the time, Fugazi did well to feel at home there. More than once, the band sat silently at some provincial service station restaurant to not risk blending into crowds of American tourists who would flood in from tour buses. They took great pride whenever European fans, especially French people, would say "You're just like one of us."

"That's the thing," MacKaye reiterated. "We grew up in Washington, DC, which is the center of United States tourism, so tourism was like a real anathema."

The collective dedication of Fugazi to absorb local culture and fit into the quotidian landscape was not lost on their growing legion of French fans. Because they were one of few bands from DC to make it to France at the time, they represented their hometown well. Their actions and engagement with French fans reflected well upon French impressions of DC.

"Ian MacKaye sounds to me like more European, [a] more enlightened guy," said Natasha Herzock of the band Kimmo. "A guy who observes. Maybe to understand, and Washington – I feel this town is like this. With people smiling, but [observational]. I can feel it like this: to learn, and curiosity, but at the same time people from Washington, I don't know how to express this. [They are] curious, but [still declaring] 'we are from Washington.'"

Herzock's perspective was echoed in connections which other Parisian punk and hardcore fans have drawn. Anthropologist Benjamin Pothier, for example, has sung in hardcore bands and done artwork for some top French punk bands, including Burning Heads from his hometown of Orléans. He grew passionate about social justice and indigenous Americans via his love of hardcore. Though he is not a lifelong Paris resident and does not consider himself a Parisian, he likens Fugazi's music to an encapsulation of those radical yet grounded politics bound up in their cities.

"[Fugazi is] very urban music," he said. "It's the music of the city, but you've got that kind of... I don't know what, even nostalgia, something like this you can feel. It's not even rock 'n' roll or about punk, it's something more poetical. You have hints of poetry and political activism, and DIY, and I think it fits pretty well with the French perspective of what art can be."

Though MacKaye was outspoken about the band's dedication to blending in wherever their tours brought them, the band's secret weapon in the French-speaking world was their other, more explosive vocalist and guitarist. Guy Picciotto's upbringing was a quintessential by-product of Washington, DC, demography and civic history. His father was born to a French-speaking family of Italian origin who lived in Aleppo, Syria, for many years. His father moved to Lebanon as a teenager and then to Paris to attend university. He emigrated to Washington, DC, for graduate school in the late 1950s. There, he met an American woman, fell in love, and decided to stick around. Guy was born there in 1965, and grew up attending the Georgetown Day School and then Georgetown University as the nascent hardcore scene coalesced in the surrounding neighborhood.

Between 1983 and 1986, Picciotto helmed the bands Insurrection, Rites of Spring, Happy Go Licky, and One Last Wish, the latter three nearly universally cited in conversations about the genesis of emotional hardcore/emo. In 1987, MacKaye convinced Picciotto to join his new, then-unnamed project. Guy embraced his role as the group's backup singer and mascot of sorts, dancing animatedly around the stage while MacKaye, Lally, and Canty jammed. At Fugazi's first gig in Paris, Picciotto shared vocals and implored the crowd, in French, to calm down so the mic wouldn't keep drilling him in the mouth while he sang.

According to varying accounts, Philippe Roizès and Manu Casana had implored the band to make a last-minute decision to go to Paris and play at Club Gibus, their only French date on that 1988 tour. A video filmed by Philippe Roger shows a packed house for their Tuesday, November 22nd, show. Local industrial metal band Treponem Pal and punk upstarts Krüll (which included Roman Jaskowski at the time) opened the show. The flyer, at MacKaye's insistence, included the names of all four members of the band. In France, like most elsewhere Fugazi played in Europe, promoters wanted to lean on Minor Threat's recognition among punk fans to help promote the show. MacKaye didn't want people to show up expecting another Minor Threat or for the show to be about him, so Roizès and Casana included "Guy & Brendan (Ex-Rites of Spring)" along with "+ Joe" to reference Lally, the band's youngest member who had no recognizable former projects.

"I recall [Fugazi's] first couple of shows there were a bit tough but that could just have been because of unfamiliarity with the group and bad luck with a couple of venues," Picciotto recounted. "By the time we'd been back a couple of times it really changed and the reaction and support were amazing as it was throughout France. For me being in a 'foreign' country where I could understand the language made a huge difference to me. Being able to communicate and interact with the crowd on equal footing were really important to us and in countries where we had a language block that was made much more difficult. In France I could follow what was going on and I could interpret for the rest of the group."

Elsewhere in France through the following decade, Picciotto would often assume duties as the de facto front man. Theatre Barbey, in Bordeaux, became one of the band's favorite venues in Europe. They played successful shows there in 1992, 1995, and 1999; recordings reveal MacKaye being uncharacteristically quiet between songs while Picciotto engages the crowd in French, despite how many of their fans could understand English. This gesture did not go unappreciated. Fab Le Roux (Thrashington DC, Syndrome 81), who saw Fugazi perform while studying abroad in Brighton, England, remarked how much he and his friends would appreciate when English-speaking bands would make an effort to even memorize and recite one or two phrases.

DC, Paris, and Franco-American Political Legacy

Fugazi's name is a tongue-in-cheek reference to the Vietnam War, the federal arms of DC and Paris' greatest collaborative failure. MacKaye, Picciotto, Lally, and Canty, all born between 1960 and 1965, came up in an America growing exhausted with its overreaching military and the veritable lost generation it was creating out of the baby boomers. The members lived and went to school mere miles from where Presidents Johnson and Nixon signed off on fateful decisions that resulted in the death and traumatization of millions. Even to the kids who would later become punks in a city unknown for its musical counterculture, political resistance was closely tied to music.

"The Vietnam War was central in my consciousness as a child," said MacKaye. "My parents were anti-war, we had a lot of anti-war protestors who would stay at our house. And the rock n' roll world was tied in with that. So I did a lot of studying of '60s culture... [reading] Abbie Hoffman and Jerry Rubin and Emmett Grogan... I think people forget with all the underground stuff that was going on then, just how crazy and horrific and savage and pointless that war was."

MacKaye landed on the name for his new band in the summer of 1987, the by-product of a lengthy obsession with Vietnam War stories and documentaries. One book that he was reading at the time, *Nam* by Mark Baker (1981), had the word *fugazi* in the glossary defined as "a fucked-up situation." Similar were the terms *snafu* (situation normal, all fucked up) and *fubar* (fucked up beyond all recognition/repair), popularized among World War II soldiers. Surprisingly, MacKaye and the band would actually need to go to Southeast Asia to discover their name's actual entomology. On November 8, 1996, Fugazi played a show in Singapore at a club called Fire. A teenage girl named Venita and her younger brother approached the band as they broke down after their set. Her brother asked what Fugazi stands for, and before the musicians could reply that they were unsure, Venita said, "fucked up, got ambushed, zipped in." That was the first time MacKaye had ever heard it spelled out, and the band had existed for almost a decade.

In Paris as well as most world cities where they played their 1000-plus concerts, they represented DC in no uncertain terms. MacKaye traditionally began every Fugazi set the same way: "Hi, we are Fugazi from Washington, D.C." Though Fugazi would speak for disaffected members of Generation X, performing several now-legendary free anti-war concerts in their hometown, they were slightly older than that age group. The members of Fugazi were unquestionably a product of their hometown, and because they all remained there for most of their existence as a band, their hometown kept shaping them. Their January 12, 1991, concert in Lafayette Park adjacent to the White House was a landmark in a media-minimized protest movement against the Iraq War. While Fugazi had played their final show before the next Bush declared the next Iraq War in 2003, the band's members continued publicly criticizing the military-industrial complex and the US invasion, both MacKaye as one half of The Evens and Joe Lally as a solo artist.

While Fugazi had earned worldwide retroactive renown for Dischord Records as genuine folk label of DC, the band's local impact and importance could not be overstated. MacKaye prefaced one outdoor concert at the Sylvan Theater in 1995 discussing how many nonprofit groups in the DC area had contacted them, desperate for help.[1]

> Every couple days or every couple of weeks, we get calls from people who are looking for us to do a concert to raise money for their particular organization, or shelter or group or something. Washington is full of social or service-oriented groups or organizations, and we're a band that has played many, many benefit concerts for groups that we think are important and fortunate for the community that we live in, which is right here in Washington, DC. … The last year or so, we've been getting inundated with calls saying 'we desperately need money,' and they need money because the funding has been cut off by the city. Because as you may or may not know, the financial situation here is really fucked. … There's far too many groups who need money, so we decided instead to just to this one for now which is

---

[1] A full video of this Sylvan Theater show is available at https://youtu.be/gZornvqsPxw.

to encourage everybody here and anybody else who happens to come across to volunteer your time, your money, and your services to these groups. They all definitely need money and the city is not in a position to take care of them, so it's up to the people. Good luck, everybody, good luck.

The city had stricken so much funding from shelters that the only financial hope came from the goodwill of a band. Congress and the DC Council had been at war over civic laws and regulations since the city had first been granted self-governance two decades prior. In 1995, around the time of that concert, Congress had stripped the city's budgetary sentience completely. Concurrently, The District's isolation from its suburbs became increasingly defined as unregulated regional development redefined and reified economic and racial segregation. This moment captured a grim reality that the city was facing, and this is worth remembering when questioning whether punk matters.

The "DC Sound" Goes Global, 1984–1994

Interviewer: Right now, where would you go if you could go anywhere?
  Skeeter Thompson: Paris (*Pit* No. 2, 1987)

Paris is an amazing city and one of my favorite places in the world independent of my experiences as a musician... it's beautiful, a great walking city, great diversity with culture and history everywhere as well as wonderful people. Very guide-book attitude but true nonetheless. Guy Picciotto (Personal Correspondence)

On December 12, 1989, Fugazi played their second show in Paris at Forum de Grenelle, an unassuming venue in the 15th Arrondissement. This came roughly one year after their Paris debut, that hastily thrown together bill at Gibus one year prior with Treponem Pal and Krüll. This time their show took place a short kilometer from the Eiffel Tower, but symbolically, the event could not have been farther from the touristic purview of the city. Despite Fugazi's paradigmatic encapsulation of the "ideal" American underground band, their impressive initial discography and relentless touring over the previous two years had attracted additional attention from an industry digging for the next big thing.

présentent

FUGAZI *G.I. LOVE*

APOLOGIZE

le lundi II dec. 89 à 19h   Forum de grenelle
5, rue de la Croix Nivert 75015 Paris M´ Cambronne

N° ·569

Droits de location en sus

Ticket for Fugazi's second Paris show on December 11, 1989

The Grenelle show sat at the end of a 26-date Fall tour that spanned the Netherlands, England, Ireland, Northern Ireland, Scotland, and Germany. Roizès, who had organized this show as he had the November 1988 show, booked the opening slot for his new band Apologize (which also included Roman Jaskowski, the only French musician to my knowledge to open for Fugazi with two different bands). Before the show began, a couple of men approached Roizès, having been directed to him when they asked who was in charge. They told him they were representing a record label in London and were interested in introducing themselves to Fugazi. Roizès did not know what it meant, but he decided to deliver the message. He walked to a small room off a corridor behind the stage where the four quiet Americans were sitting and reading. Roizès recalls opening the door, and in a creepy synchronicity, MacKaye, Picciotto, Canty, and Lally all looked up from their books. Philippe told Ian that a couple of label men were out front and wished to speak with him. Ian told Philippe to let the men know he appreciated that they came out, that he hoped they enjoyed the show, but there was nothing to talk about.

The Forum de Grenelle show would not be the last time that the band would confront major-label attention. The 1989 anecdote feels like presage to the major-label feeding frenzy that would ensue in 1991 leading up to the Grunge explosion. The pressure that Fugazi faced to operate within the major-label universe is a focus of Joe Gross' *33 1/3* volume on 1993's *In on the Kill Taker*, the band's best-selling release and high-water mark of their mainstream popularity. Though Fugazi attracted the curiosity of the mass media, they democratically maintained their decision to continue operating independently. As Sarah Thornton (1996) wrote, "more than anything else, the underground define themselves against the mass media… [their main antagonist] who continually threaten to release their knowledges to others" (p. 117).

Few better examples exist of this convergence between the corporate mainstream and the underground than Ahmet Ertegun's early-1990s pursuit of the band. According to Picciotto, the Atlantic Records chief showed up backstage after a Fugazi gig in New York. Though the band would have rather just heard stories about bands they loved from Atlantic's history, they respectfully listened to his pitch with no intention of taking it too seriously. Ertegun wrongly assumed that Atlantic was the first label to make the band an offer. A&R people from several labels were aware of the members' past projects, so when Fugazi began touring in 1987, they carried a "DC super group" tag among indie aficionados. In 1993, Ertegun had succeeded in signing Jawbox, who were the first band to leave Dischord for a major, setting a precedent for Shudder to Think to do the same with epic soon thereafter. Fugazi, however, had no realistic reason to budge. Dischord, due to its sustainable business model, was never suffering financially. This reality contradicted some French punks' impressions of it, which romanticized the DC bands they were discovering around that time. That being said, many French fans who got into the band on their later tours considered Fugazi peers of American indie rock heavyweights.

"I think [Fugazi] were seen as one of the major US acts, the way that Sonic Youth were," said Paris scene veteran Gaël Dauvillier. "The difference might be that one was more outspoken than the other. Bands like Prohibition and alternative bands from France were definitely influenced by Fugazi and they were spreading the word. My friends were doing distros and it seemed super easy for them to get their records for cheap. When I was young I remember them saying, 'You should listen to Rites of Spring and Circus Lupus and Slant 6 and the Make-Up' and all that. We were so excited. Maybe people and the [90's indie] press were just seeing them as like Sonic Youth, Pavement, or bands like that from that era."

As punk's success globalized, so did its corporate consolidation. Somewhat ironically, punk's corporatization was integral to its globalization. Many less developed countries like Indonesia would not have thriving punk scenes today if Green Day hadn't released major-label records that had well-marked conduits for international distribution. For several reasons, Fugazi became de facto ambassadors for DC punk; they toured more and existed longer than most any of their contemporaries. Scream and Government Issue both technically existed longer than Fugazi, but only in name. Scream's classic core lineup (Franz Stahl, Peter Stahl, and Skeeter Thompson) still performs on occasion, and Government Issue never technically ended until singer Jon Stabb's death in May 2016. Though Scream beat Fugazi to France, performing in Paris in 1986 with Sherwood Pogo, they were not quite as prolific or successful. Fugazi played over 1000 concerts across almost 40 countries on 5 continents between 1987 and 2002. The band recorded a majority of these performances, over 800 of which have been remastered and made publicly accessible on the Dischord website.

Fugazi were the most iconic, best-selling, and spiritual leaders of DC punk and, in the eyes of many critics, still the standard-bearers for punk's artistic viability. That Dischord's label owner was one of the four members had some bearing on the band's role as spiritual ambassadors, but Fugazi might have succeeded as a major label band. That being said, it was best left to mystery, considering how they still represent a platonic ideal of successful independent musicians to many international fans. In turn, their hometown has become something of a platonic ideal for DIY culture and anti-corporate resistance.

Fabrice and Nicolas Laureau discovered harDCore in 1988, shortly before they began practicing and playing out an early iteration of Prohibition. They had kept skateboarding since learning to do so in Chevy Chase, Maryland, as little kids, except now they were listening to Minor Threat and Bad Brains while skating around Paris. Their family had also lived in Nigeria for a couple of years, so Fab and Nico were both Afro-beat fans long before most Westerners got into it. As lovers of world music in general on top of rap, funk, hardcore, and punk, Prohibition made liberal use of saxophone and sitar. The Laureaus did not discover Fugazi until 1990, but became huge fans as soon as they heard *Repeater* and the first two EPs.

"I think we had kind of the same approach and same feeling in the wish to mix a wide range of sounds and grooves as other bands including some DC bands," said Fabrice.

Prohibition indeed found worthy counterparts and kindred spirits who they would support for dates in Angers, Toulouse, Bordeaux, Marseille, and Lyon (the latter show produced by Maïe Perraud) in Spring 1995. By that time, the band had gotten Prohibited records off the ground, a label that owed no small debt to the Dischord model.

"We were not the only ones in our generation, but we invented something compatible with [certain] French expectations and rules," wrote Nico, citing the value of "culture" in France, "[as well as] what we learnt from the American punk scene and more specifically Dischord, but also the K Records or Touch & Go experiences."

By 1996, Prohibition had also helped found Push, a collective practice space that enabled a litany of experimental groups rehearse within Paris' restrictive confines. The LP the group released in the wake of starting the label, 1996's excellent *Towncrier*, sought to musically represent and refract life in Paris. They saw how many of their favorite DC bands like Fugazi heavily represented their hometown lyrically and graphically (e.g. the Washington monument on the cover of *Kill Taker*). Though it would be reductive and unfair to call Prohibition the French Fugazi (or vice versa), several Parisian punk fans have implied the parallels between the two bands' innovative designs on punk and DIY. More importantly, both bands accomplished much in representing the artistic establishments and spirits of their respective cities. Reflecting on their relationship and the string of shows they played with Fugazi in 1995, Nico adds, "They gave a reality to our identification, and [became] sorts of mentors; they suddenly became gentle big brothers, if I can say so."

Bibliography

Thornton, S. (1996). *Club cultures: Music, media, and subcultural capital.* Middletown: Wesleyan University Press.

Pit #2, 1987 (DC)

7

Earthquakes Come Home: French Punks Visit DC

In 1984, two years after unleashing Bad Brains on the music-consuming public, the label ROIR released its *World Class Punk* compilation, among the first of its kind to unite bands from 25 countries on one easily traded piece of vinyl. Kevin Dunn (2016a, b) called the collection "an important reflection of, and introduction to, the increasingly global scope of punk at that time" (p. 134). Notably, it introduced the Parisian hardcore sound to the world with the song "Detourement" by Nevrose, a mysterious band that only existed for one 7-inch release.

Given the mythical following that Fugazi and the DC scene at large have accumulated in France, it is not surprising to find French punk fans in the District on their American holidays. Their anecdotes from the trail of spaces and places of harDCore underscore multiple geographic ideals, not to mention the changing nature of tourism in the twenty-first century. These include the discussion and "meaning" of tourism and place heritage, the spectrum of "clean" versus "dirty" tourism, social space in the context of travel, and how underground tourism ultimately contributes a radical new value to otherwise unremarkable places.

Washington, DC, receives a significant number of French visitors every year, many of whom come to town, like many Americans, during formative teenage years. Many French students who would grow to love

DC punk and figure prominently into this story visited the city on holiday well before they had an idea about what was happening underground. Such was the case in 1982 for both Maïe Perraud of Lyon and Craig Wedren of Cleveland.

Perraud, 16 at the time, was a French boarding school student on her first trip to the US as part of a cultural exchange program. Her group's visit focused on New York, but they included a side trip to DC. Perraud remembers listening to punk at the time, but wasn't familiar with Minor Threat or Government Issue until years later. In 1990, and 1995, her collective Gougnaf Movement would organize Fugazi's shows back in Lyon.

Wedren, who was 13 and on his eighth-grade class trip in 1982, would move to DC with his father three years later during the zeitgeist of Revolution Summer. He enrolled in the Field School in Georgetown, and within a year was playing in the band that would become Shudder to Think. "If I had known what was going on [in DC in 1982], I would have hopped right off that tour bus," he laughed. Shudder to Think would play in Paris, Orléans, Poitiers, and Lyon in May 1990 on their first tour overseas.

Though DC scene reports and harDCore recordings were circulating in France and playing on shows like Radio Bellevue's *Western Front* as early as 1981, few French punks had the resources to dig deep enough into *l'Amerique profonde* and get much farther than New York or Los Angeles. Perraud cites the weekly TV show *Musique California* as one of few mainstream outlets for American punk in France at the time, but that show (aptly) focused on bands from Los Angeles and San Francisco. The first two American bands she saw were the Los Angeles new wave group Wall of Voodoo and the Sacramento garage legends The Cramps. Though she still holds DC in high regard musically, she has never returned, instead following her network to San Francisco at least eight times in her adult life. She recalls first reading about Dead Kennedys in *Best* magazine, which quickly led her to discovering Alternative Tentacles; Jello Biafra's label turned San Francisco into a cultural destination for her in that way that Dischord had for DC. After mail-ordering her first Government Issue Record, she discovered Minor Threat, which led her into contact with Stéphane

Cressard, the record-trading bassist for Flitox in Paris. Maïe soon thereafter "bought a subscription to *Maximumrocknroll*, and then it was over," she laughed.

Long seduced by the lure of the "real" America, French music lovers had been traveling to and catching musical performances all over the US for as long as the country had existed. Divining who the first French visitor was to experience the American punk underground via any specific show would be impossible, but there is evidence that Gérard Miltzine of Grenoble attended hardcore shows in DC on holiday as early as 1982. These cases were rare at the time, as the only outsiders who knew about what was happening were those closely associated with the bands and Dischord. Wedren, for one, is skeptical.

"There was no way that any tourist would have known to come to DC Space or a house show that was happening in the early 80's – it just was so far under the radar," he said. "I imagine it was probably the same in France. The music scene in France was so weird … My impression [at the time Shudder to Think started touring] was that basically before the French house scene started happening, there was virtually nothing happening… we really didn't know, when we toured there, where to go [or] what to do. Where's the music? Where's the new thing happening? I mean, there's the music that's been unearthed over the past 30 years… but I never got the impression of any kind of cosmopolitan scene, whereas in [other countries like Germany] there was something going on."

By many accounts, the first Parisian punk to visit DC specifically motivated by harDCore was Manu Casana. In 1984, he was singing for Sherwood Pogo, running a fanzine called *Dekapsuleur*, and managing a small record shop on Rue Véron called Terminal Records. He set out in April of that year to establish contacts with American indie labels, seeking artists to import and eventually to help other French indie distributors. He landed in Los Angeles in mid-April, spending ten weeks in the US. Manu spent three weeks in Los Angeles, two in San Francisco, three days in Seattle, three days in Chicago, and two and a half weeks in DC. He would return to Paris from New York at the end of June 1984.

Manu Casana, photo/illustration by Tomas Squip. (Courtesy of Dischord Records)

Like many punk-influenced visitors, Casana stayed at the Dischord House in Arlington. Tomas Squip, Beefeater's vocalist, took a photo of Manu that he doctored and hung on the wall of the house; it stayed up there for years. Even in the early 1980s, well before Scream and Fugazi went overseas, the residents became accustomed to international visitors.

"[Before I ever went to France,] I had met French people," said Ian MacKaye. "I had met Manu, I had met Philippe [Roizès] and Arnaud [Gabelli], but I knew them as Europeans. International travelers stayed at Dischord House all the time… at one point I remember we had two Germans, a Norwegian, a Swiss woman, and a Dutch woman all staying at the house."

Most touring bands and friends of the label would usually be around for a few nights before heading back on tour. Casana stayed for 16 days. His prodigious height and radical accoutrements made him stand out, even among the Dischord circles. Whenever he left the house for the day, he always wore a handkerchief and a pair of goggles around his neck. "When I asked him why he always wore them, he said, 'In case I get caught in a riot!' to protect him from tear gas or something," recalled MacKaye.

Ian MacKaye, with Manu Casana (on the couch) at Beecher Street, Spring 1984

Anecdotes of Casana's time in DC read like a sitcom script about a visit from a wacky foreign cousin. More than once, he left in the morning for standard tourist activities like a visit to the Smithsonian museums, and then called the Dischord House late in the evening from a payphone, lost, miles away from his intended destination. He once encountered a small crowd in a tent near the National Mall, the smell of food having drawn him inside. Within minutes, he was eating free food and chatting with members of the group, helping himself to generous servings. He had assumed it was a socialist barbecue like those to which he was accustomed in Paris, where local workers and less fortunate had an open invitation to stop by and enjoy a meal. This, however, was a family reunion; the family in question found the gregarious French twentysomething entertaining in his getup and broken English, so they let him stay and invited him to join in various photo ops. Within days of that picnic, he got lost after joining a group of people who appeared to be "running from something" across Memorial Bridge in the direction of Arlington National Cemetery.

He figured he would run along in order find safety from whatever was chasing them. Clearly, jogging was not a popular activity back in Paris, or at least within Casana's circles. Hanging around the house in Arlington, MacKaye also remembered Manu being "blown away by these little creatures; he couldn't stop looking at them." He referred to them as "little long-tailed monkeys," which gave Ian the impression that there were no squirrels in Paris, either.

The Dischord crew became accustomed to Casana's quirks. The trip ultimately bore fruit for his business ventures, too, helping lay a groundwork for a greater international record distribution in France. In June of 1986, Manu wrote to Ian to follow up about his progress with starting a new label and his attempts to get Dischord Records distributed in France. Also, he wrote extensively thanking him, the MacKaye Family, and the Dischord Family:

> *Maybe it's not the same for you, but I really appreciated my time in W.D.C. And must give you many thanx [sic], you don't feel it, but I feel that… cause what you said was and is right. I was thinking the same but I never push my ideas like you, I never practice like you cause I never get the possibility and the force to do it, but now I care more and the important [thing] is to do what I feel. And it's another moment to move, because the times are really hard now in France, fucking government who impose [their] police everywhere. You can't go to the subway without a police controle and it's a shit time too cause they close many ballrooms and others… fucking ripoff.*

Sherwood (Pogo) put out one full-length LP *Liberté* (1986) on his new label Autodafé before dissolving. He named his new distribution Plus au Sud (*More Southern*) a nod to John Loder's label, and he began to distribute Minor Threat and Government Issue records in Europe around that time. At the beginning of 1988, he wrote back to request 100 copies of the first Soul Side LP *Less Deep Inside Keeps* and 50 copies of the *Four Old 7"s on a 12"* compilation LP. Within a few years, Manu would transition to promotion and production for rave, another underground genre that fetishized desiccated North American and British urban landscapes. Though it has been decades, Casana's legacy as a Parisian hardcore and punk pioneer, at least in the sense of his experience as an unconventional music tourist can be understood, keeps his name in conversations about the history of this circulation.

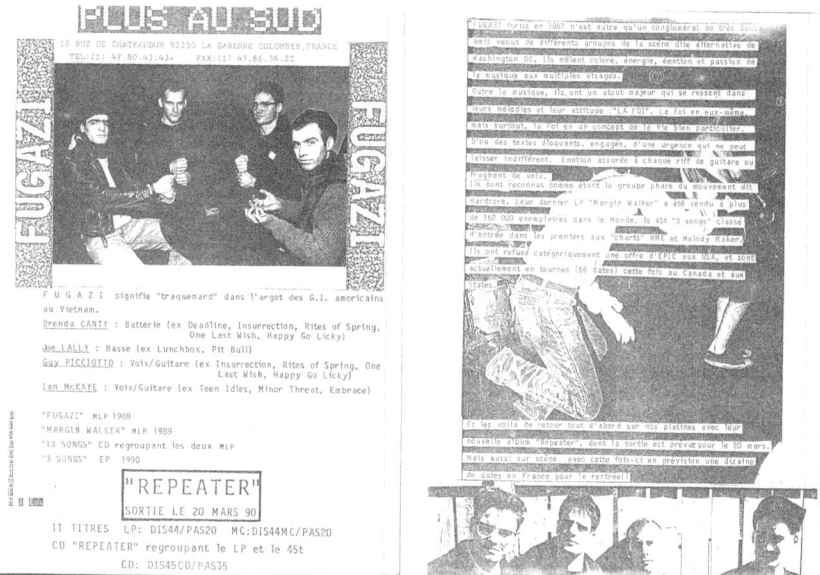

Plus Au Sud promotional insert for the French release of Fugazi's 1990 LP "Repeater"

Meanwhile, the Paris cadre of hardcore fans would meet every Saturday at Café Verdun, a dive around the corner from the Gare l'Est train station. The group, which often included Roizès and Gabelli, would meet at noon to trade tapes, share ideas, gossip about the scene, and, to the chagrin of management, nurse their drinks for several hours. The ones who occasionally escaped Paris to explore greener punk pastures saw their statures elevated, their stories eagerly anticipated upon their return.

In July 1987, Roizès and Gabelli, both 20 at the time, paid an extended visit to Dischord while on holiday in the US. Roizès recalls that he expected DC to be more "lively" than it was, but everybody he met was nice and welcoming. In particular, MacKaye took care to speak slowly and repeating himself when necessary for his Parisian visitors to understand him. Their conversations about music and politics would last for hours into the night. Philippe snapped some photos of early Fugazi practices in the cramped basement, scant extant documentation of the band before Picciotto joined. The following Fall, Roizès followed up with MacKaye, writing him with updates on the French bands who had split up, the new developments on the distribution service he ran with Casana (by then called MD Diffusion), and mentioned plans to contact Don Zientara about Flitox wanting to record at Inner Ear Studios on their own US visit a few months later.

Arnaud Gabelli with Mudpot, Dischord House, Summer 1987. (Image Courtesy of Philippe Roizès)

Arnaud Gabelli and Philippe Roizès hanging out in Dischord's backyard, Summer 1987. (Photo by Jeff Nelson, Image Courtesy of Philippe Roizès)

Ian MacKaye sings at an early Fugazi practice in the basement at Dischord House, Arlington, Virginia, July 8, 1987. (Photo by Philippe Roizès)

DC's Political Landscape as Punk Landscape

Though the punk scene's self-determination relied somewhat upon their opposition to the dominant federal government, Washington's status was a magnet for an international crowd of activists, service workers, and politicos brought so many denizens into the scene in the first place. Many people pictured and quoted in the books like *Banned in DC* attributed their DC roots to the government, including Kenny Inouye, whose father David served as a US Representative and Senator from Hawaii until his death in 2012. "It was always so exciting. You had the feeling that when you went to the show you were in the middle of something really big. Something was really happening," said Inouye about his experiences first going to shows in spring of 1981. He would eventually found the quintet Marginal Man, one of few local bands to leave the mid-Atlantic and tour routinely.

The irony was never lost on Cynthia Connolly that the ascension of an ultraconservative president and administration was a major catalyst for the underground movement in 1981. After her mother moved her and her sister Anna into Hawthorne Street, they began to enjoy proximity to the epicenter of the emerging hardcore punk culture, two miles north of Georgetown's core at M and Wisconsin. While perhaps the flagship of DC's most intense gentrification, the neighborhood was a pivotal center for the performance and networking of quotidian punk culture throughout the 1980s and even a substantial portion of the 1990s. In the first issue of her zine *Capitol Crisis* in 1980, DC punk trailblazer Xyra reviewed a Teen Idles set opening for veteran rockers S.V.T. at the new 9:30 Club:

> Between songs, the T.I.'s and their fans chanted "Georgetown Punks" repeatedly, which meant nothing to anyone else. I found it hysterical, sad & confusing, as the last place in the world you'd ever expect apparently rebellious youth to identify with would be the Posh shopping district of Georgetown! I guess they all live around there and it's their way of thinking they count, as a few colorful characters, dotting the busied street.

As Mark Haggerty, the guitarist whose resume includes Iron Cross, Gray Matter, and Three, told Connolly, "You'd start at the top of Georgetown and you'd go to Haagen-Dazs, say hello to people, go to Crumpets, get day-old free stuff, say hello to people, then you'd walk to Swenson's, say hello to Danny [Ingram of Youth Brigade], then you'd go to Record and Tape. It was just like, say hello to people and get some free stuff."

Many punks would retain jobs in the neighborhood through the end of the decade. When Roizès and Gabelli first visited from Paris in 1987, they spent a lot of their time in that area. Most of the activities in which they engaged happened within what felt like a one-mile radius of Georgetown. The social landscape of DC punk presented a stark counter-narrative their preconceived notions about the District.

Although most of the record shops like Record and Tape, Penguin Feather, and Orpheus either closed or moved to cheaper areas by the end of the century, Smash! held on to their location near M and 34th until 2006, when rents and a changing culture in Georgetown sapped their venture of any sustainability. Smash! would relocate to the Adams-Morgan neighborhood, which by 2010 became a cluster of record shops that also included Crooked Beat, Red Onion, and Joint Custody. While they were hardly carbon copies of one another, all four paid a degree of homage to harDCore. Crooked Beat has since relocated to Alexandria, Virginia, due to persistent problems with rats in their building. The shop's owner Bill Daly told me that contractors estimated that the construction of a new hotel north of Dupont Circle displaced over 100,000 rats, many of which took up residence in the older, vulnerable buildings along 18th Street. Joint Custody and Red Onion, two other street-level shops a few blocks up 18th, have also relocated to nearby U Street as of this writing.

The first generation of Parisian hardcore punks to visit DC, particularly Casana and Mension, were thrilled to find a community that was as politically charged as it was musically innovative. Minor Threat, Scream, and other bands in the Dischord family grew up in the shadow of the war machine that had already inspired generations of protest music. This tense dynamic equally intrigued and confused French fans who came to discover the scene and its hometown.

"My [American] friends were [emailing] me about [Iraq War protests in 2003], and I thought there was definitely repression here [in France], but nothing comparable," reflected Gaël Dauvillier. "I remember them telling me there were these protests where people were riding bikes, and during the whole protest there was a helicopter flying over them, crazy things."

Dauvillier could not have imagined something like that happening in France in the early 2000s or even now. Coming from a place where the only people he saw fly the French flag were likely nationalist and far-right, it came as a culture shock to visit a friend's parents' house in the US and seeing American flags everywhere. As someone whose parents were heavily involved in the 1968 student protests in Paris, Dauvillier was intrigued by the DC scene's devotion to political action.

"In France, everyone knows of, or has a family member that is radical in some way," he stated. "I realize that Fugazi were taking stances against wars led by the US, and imperialism, and these were progressive ideas and this was something very inspiring to me. And I heard about the DC Women's Rights rally, and about the connections between DC and Bikini Kill and Riot Grrrl, and I thought, 'Damn, this is really what I am into.'"

Politically, much like their Parisian counterparts, the denizens of DC hardcore were a mixed bag. The city's legacy as a progressive hotbed won out, but many kids in the scene, particularly those attracted by the band Iron Cross circa 1982–1983, supported right-wing causes and administrations. Although most of the DC punk scene came from middle-class backgrounds, the Positive Force community understood Ralph Knupp's (1981) maxim that "problems are more rhetorically fruitful for movements than solutions" (p. 383). They prized community and grassroots political action as forceful statements and "beginning at home" idealism. The year 1985 marked "Revolution Summer" as the hardcore kids came of age, mixing pop with politics in a socially constructive way.

Though the DC punks confronted social injustice domestically, they also used their resources to confront international injustices like Apartheid. Rock stars like Steven Van Zandt, Bruce Springsteen's guitarist and eventual *The Sopranos* star, organized an all-star cast of musicians to declare a boycott of concerts sanctioned by the South African government in "Sun City." The mid-1980s became the golden age of all-star pop singles, between Artists United Against Apartheid, USA for Africa's Lionel Richie and Michael Jackson–penned "We Are the World," and Bob Geldof's Band-Aid single "Do They Know It's Christmas?" But way off of the pop music radar, the DC punks realized that, unlike superstars like Bono and George Michael, they had the ability to take to the streets and create disruption in their own backyard.

The South African Embassy sits on Embassy Row on Massachusetts Avenue NW, a mile north of Georgetown, a mile northwest of Dupont Circle, and a few blocks from the then vice president Bush's home at the Naval Observatory. Punk percussion protests became common across the street from the Embassy by late 1985, continuing a longtime international protest tradition that continues today globally. The musicians who helped "liven up" the protests in 1985, including Rites of Spring, fit this description. As Picciotto said at the time, "I'm not going to try and propose any solutions. I'm just vocalizing my rage, that's all." Picciotto, along with band mates Eddie Janney and Brendan Canty, all around 20 years old at the time, related the genesis of their band's involvement in the Embassy protests to the zine *Truly Needy* that year:

Picciotto: [The protesters already there] would show up every day and do the same march. We thought we'd inject a little spontaneity into it.
Janney: It was really cool. About 50 people showed up…
Picciotto: With drums, beating the shit out [of] them in front of the Embassy…
Canty: … for two hours straight.
Picciotto: We screamed 'Freedom, yes, apartheid, no!'
Canty: The people there really dug it. There was a guy playing drums. They invited us over and kind of welcomed us into their own thing.

The punks' involvement in the anti-apartheid protests became a uniting event, transforming an otherwise unremarkable stretch of grass next to the road into a site of historic political activity. A sizeable cadre of local musicians, fans, and friends filed to the Embassy after school and work in the afternoon. It also changed the acoustic signature of a massive radius of Northwest DC, which was filled with many affiliates of the Reagan administration. Mike Hampton, then the guitarist in Embrace, could hear the thumping from his house in Georgetown, a mile away through Rock Creek Park.

To the average tourist, the South African Embassy may not have been a crucial stop on a sightseeing tour, but the fact that South Africa's government was long responsible for such a crime against humanity created a disruption that manifested itself musically. Like the *Cacerolazos* drum protests around the Argentine election in Buenos Aires in 2001, these exhibitions impacted and transformed urban spaces. The protests created a disruption that visiting punks from Paris were enthusiastic to join and informally unite their home scene with this exciting new one they were approaching. Not only did DC provide this opportunity for participation, but it also gave them the microcosmic platform for grass-roots political action in the world's most powerful city. It gave these alternative tourists a mainstream sense of belonging. One of them, Norbert Mension, came to DC around the time that Heimat-Los broke up in 1988.

"When I was in DC in 1988, there [was] a big march for peace," he recalled. "Reverend Jesse Jackson, who was a candidate for the Democratic presidential nomination, participated [in] it. And I think the same day or a few days later, Positive Force, a punk activist collective from DC, organized one of these protests against apartheid. I was staying [with] Mark Andersen from Positive Force. Most of the musicians and music activists participated. There were people from Dischord Records. We all brought percussion instruments and we had to hit them very loudly just in front of the South African embassy to disturb people working inside. This protest existed from 1985 to 1990. They had this concrete involvement making a link between music and politics."

Though thousands of miles separated Washington from cities like Johannesburg, civic issues in the District bore many similar scars of institutionalized racism. It was no secret that black residents in both cities lived in disproportionately poor conditions, much more prone than their white counterparts to the rapidly intensifying AIDS crisis. One flyer for an early 1990 gig Fugazi played with Fidelity Jones and Lucy Brown at St. Augustine's School advertised it as a benefit for Positive Force "against apartheid here and in South Africa."

Other visitors from Paris in that span also found their way into the sea of banging drums and pots on Massachusetts Avenue. Stories like these, of immersive punk experiences, circulated back to France with the visiting Parisians who experienced them. When Philippe Roizès and Arnaud Gabelli got back from their extended trip in 1987, they felt like celebrities among the group of punks who would gather at Café Verdun every Saturday. Roman Jaskowski, then the bassist for Apologize and Krüll, recalls those Saturdays fondly. He remembers enjoying his friends' stories about this enigmatic city before he and his band mates left to go meet up with some metalheads at the FNAC in Montparnasse.

"All the impressions [I had of DC] were through stories that were told by Philippe, who already had spent so much time in the Dischord House, and he had so [many] cool stories, so when I went there, I already felt like I knew everyone," said Jaskowski.

In September 1990, Roman arrived in Washington, DC, for the first time and felt an immediate culture shock, having immense difficulty communicating in English with airport officials. He quickly found comfort among the rapidly evolving punk scene there, in spite of stereotypes that had circulated about how rigid and straightlaced its members were.

"I was definitely not disappointed by the warmth and the openness of people," he reflected. "It felt right away like we were part of a big family. So as far as how welcoming and how funny and open discussions can be and endless. At the time I discovered the states, I didn't realize it, but there was more culture more knowledge, more wit, as far as being European, I felt that the city where I can relate to people as far as conversations was DC. So I felt real comfortable, but I'm upset to be honest that at that time I was already partying a little bit. I liked funk music and smoking and drinking and dancing. And to me, DC always had this seriousness and righteousness, which I am not saying in a negative way… it could be relaxing and wise and soft-spoken, real adult."

Roman Jaskowski (L) with members of Jawbox (L-R Kim Coletta, J. Robbins, and Bill Barbot) in Silver Spring, Summer 1994

Jaskowski's "in" with the DC crowd, outside of Philippe's vouching, was Shudder to Think, whom Krüll had opened for earlier that summer in Paris. Roman got Stuart Hill's phone number, and wound up staying at the bassist's group home in Silver Spring. During his visit, Hill would come home with a couple new mixes from his band's upcoming EP, *Funeral at the Movies*. In 1994, Roman returned to DC, spending time with the then major-label Jawbox at their house in Silver Spring. Stories like Jaskowski's call attention to how circulation is more than simply information and ideas moving between cities and friend groups; a large component of it is an emotional geography of belonging.

The Dischord House as Alternative Tourist Destination

The Dischord House and other sites affiliated with punk inform French punks' mental maps of the DC landscape as much as any museums or monuments. Where most research on specific scenes has focused on the musicians and venues, indie labels themselves have received relatively little attention. Because many of these labels have been mere bedroom operations, most proprietors have laid bare their mechanics. Where the major labels were conducting business behind closed doors, the indies prided themselves on accessibility. Where big business closely guarded the details of their deals, the underground prided itself on disclosure. Manchurian pop-punk godfathers Buzzcocks diagrammed the expenses of their first 7-inch *Spiral Scratch*, a radical move when they released it in early 1977. Though tiny jazz, country, and blues labels released music to committed small audiences for decades before punk or even rock 'n' roll, independent labels have embraced multiple methods of disalienation. Effectively, this meant keeping themselves transparent and more accessible to their fans. This manifested in the form of personal home addresses, emails, and phone numbers included with their products.

This transparency, in pinpointing sense of urban place for visiting music fans, has had both positive and negative consequences for labels worldwide. According to Michael White's biography of Bristol-based boutique label Sarah, Anglophile indie pop fans from around the world started showing up regularly at their flat on Upper Belgrave Road. Cofounders Clare Wadd and Matt Haynes were flattered by these pop pilgrims, but it started to become overwhelming considering how long their workdays already were at the peak of the label's activity, without having to humor a stranger for an hour or more.

Dischord experienced a similar phenomenon early in the label's existence. Beecher Street NW, an unremarkable residential row off Tunlaw Road in Glover Park about a half-mile from the US Naval Observatory, started drawing curious punk fans from far reaches of the globe. Even those from the neighborhood, including a teenage Stuart Hill, were in similar awe of the MacKaye family home over the 1980s. Today, even though most fans differentiate between the house Ian, Alec, and Amanda MacKaye grew up in and where Dischord actually conducts its day-to-day business, the Beecher Street house retains a high status in the international imaginary of the District. The family home actually adopted the nickname "Beecher Street" in its function as a clubhouse for the nascent punk movement while the MacKaye siblings still lived there.

"What I really like, as Geography is involved, the perception of the city for me is the address," said Florian Pons of the Lyon emo band Sport. "Every time I had a record [on] Dischord, there was this address written on it. And it made me want to check it, because it's really important for me – the place where things are made. You have the feeling like faraway, you can get the vibes of the thing… and that's the perception of the place for me, it's big… If you go there, you only go first to places you want to see, and what is the address, it's a written thing, you know? So you type it on your GPS or your phone…and you ask people."

Hugo Maimone, on a trip to the US with a Dance Company in the Fall of 1989, walked miles over into Glover Park in an attempt to deliver a copy of his band's LP to Ian MacKaye, finding the hallowed address in what seemed to him like "the middle of the woods." He knocked on the door and stood back, and an older woman answered. Maimone will likely never forget that moment, horrified that this woman would call the cops on him. She said hello, and he asked for Ian. Ginger MacKaye smiled and replied that she was Ian's mother, invited Hugo in for a glass of water, and got Jeff Nelson on the phone. Ironically, Fugazi was on tour in Europe. For Ginger, this was not the first time this had happened. When Ian moved out of Beecher Street in 1981, he left a letter for all of his fans who visited explaining that he grew up and ran the early iteration of Dischord Records in that house. He decided to keep his parents' home as the official address for Dischord correspondence simply because they were likely to stay put. This meant that within 18 months of Dischord's first release, the label's releases had generated enough mail to necessitate a stable mailing address that would prevent changing the address on future releases and risking returned letters and, in the case of international fan mail, dead ones. Undoubtedly, Ginger MacKaye's support for her son's project prevented him from needing to rent or purchase a P.O. Box. Circulation relies upon strategic gatekeepers and conductors, and musicians' personal and family networks were crucial to this proliferation as well. Ginger would remain a valued ambassador for the label and DC's punk scene to visitors from across the globe for decades until her death in 2004.

While the Dischord mailing address has remained static for the lifespan of that label and prominent within the mental cartography of DC for French fans like Pons, the Arlington house has provided the visual accompaniment for harDCore in the international imaginary. Like Beecher Street in Glover Park, the Dischord House's unremarkable appearance has been a strength as an alternative de facto pilgrimage site. It looks no different from any other house on the quiet Fourth Road in Arlington. Unlike recognized historical places, of which there are countless within a two-mile radius of either address, there is no official signage, no museum, and visiting unsolicited is discouraged. Perhaps this is another reason for the sustainability of the label, as all of the tourists (including unsolicited visitors) understand that they are going to experience a "real" place that has not been manicured or altered for or by the tourist gaze. Tourism scholars like Kevin Fox Gotham, who specialize in the French Quarter in New Orleans, call this dynamic the "hyperreal, by which people lose the ability to distinguish between the 'real' and 'illusion'" (Gotham 2005, 1738).

Where DC's most famous addresses gradually accrete sufficient symbolism and memorial meaning to approach "hyperreality," the city's mythologized punk landmarks remain quotidian and humble. Most of the structures that have weathered gentrification, like the Wilson Center, remain in use as a functional community center on 15th and Irving Streets NW. Those sites which have been less fortunate have closed and quickly become subsumed into the urban landscape. DC Space, the Penn Quarter club that hosted the first hardcore punk festival in late 1980, closed in 1991 and, like so many historic buildings with future foot traffic, eventually became a Starbucks. The nearby Lansburgh Center, where Minor Threat played their final show in 1983, originated as the department store Lansburgh's, which closed one decade prior. Today, furnished extended-stay luxury apartments carry the name.

These unpretentious tourist destinations throughout DC—the Dischord House(s), the 9:30 Club, Wilson Center, DC Space, and many others—follow no geographic pattern and, unlike most proper tourist destinations, are not under the control or influence of the hegemonic tourist industry. City hall and the federal government may control property taxes and zoning, but both the meaning and the counternarrative role of these places are relatively unthreatened. Perhaps the lack of historicizing or memorializing, especially for former venues that are now chain coffee places, is helpful. They make possible the preservation of the network—the virtually private, guarded ownership of the sites' meaning by a knowledgeable minority. Sometimes, these underground geographies are content to remain underground while they can.

Urban Life and Political Contrasts

While the "pull" factors, or landscapes of attraction, are understandably foregrounded in discussions of tourism, the motivating and enabling "push" factors must also be considered. Those from the most developed countries engage in the majority of tourism globally, so touristic conceptions of place cater to the imaginaries of those who can afford to visit (monetarily or time-wise). Consequently, the same economic and social classes dominate much of the discourse on tourist landscapes. Many facets of French culture and citizenship helped punk fans and musicians spend time exploring scenes outside of their home country. Those who worked full-time jobs had a well-documented abundance of leave time, which has made it easier for French bands to tour as long as the touring circuit has existed in Europe. This discrepancy between French and American leave time was a stark contrast that several of my informants mentioned during interviews, some in a tone more ridiculing than others. Leave time, however, was just one of multiple cultural-political factors which often generated a heated debate about France's social-welfare model versus the Anglo-Saxon business-welfare model.

For the life span of the Republic, France has emphasized "liberté, égalité, fraternité" down to its motto. This has permeated innumerable cultural mores, which has landed to the advantage of French musicians in many respects. Even if the National Assembly did not take punk seriously, it recognized all music as a vibrant part of culture. Especially for the 1980s generation, the cultural turn of François Mitterrand's administration has remained crucial.

"When Mitterrand was elected, the left wing is very cultural, you know, we are in France and in France there is a strong philosophical and artistic culture, and he wanted to magnify it," said Maïe Perraud. "And in the eighties all the young people, those who are now 55, at the beginning of the eighties wanted to make music and play in bands, and they could do it. And they invented a special status for 'the music workers,' for all the artistic world, there was a new 'intermittent du spectacle,' which means people who worked part-time for cultural wealth. I think that the Mitterrand years did a lot for culture and opening the doors and to let people from all over the world come here to play and to show what they were doing."

Coming from the other side, American touring musicians found that environment to be a breath of fresh air compared to a neoliberal, capitalist model that limited leave time and in general devalued artists unless they generated major income. This rang especially true for artists from a relatively "uncultured" town like DC, where they had to make their own scene.

"I was talking to someone recently who told me one of the principles when they were rebuilding Europe in the 40s and 50s, there was a lot of focus on creating culture," said Ian MacKaye. "They were putting a lot of money into culture; you could get paid for playing music. Maybe they saw it as branding in a way, but they thought, 'This was important; we need to promote our cultures.' And certainly if you think about things like healthcare – that issue in [the US] strangles creativity… If you have a job that gives you health care, you're on the fuckin' wheel, every day. But I think healthcare and living accommodations are really central. Not that it's so cheap to live in Paris, but at least you get a fuckin' band aid for free. So, I think in Paris at least culture is taken seriously, and not in purely commercial terms, so that's why I think there's always been an appreciation for American culture…. I think they kind of validate artists over here because the only way you get validated over here is to win an award…As a punk rocker, we take ourselves pretty fuckin' seriously. Americans don't take us seriously, because we're not making any money for them. But over there, they're like, 'You're legitimate.' And I think that's what most human beings want is to think they're legitimate, and what they're doing is real."

Similarly, the following generations of bands who carried the District's cultural cache to Europe were able to enjoy the spoils of the groundwork previously laid through the circulation of harDCore. The Capital City Dusters, for one, enjoyed a set of warmly received shows in central France around the millennium. One of chief Duster Alec Bourgeois' favorite shows he ever played was in a tiny avant-garde theater in St. Etienne in 2003.

"Being from DC always made a positive difference when we toured in Europe and in places like St. Etienne and Lyon specifically," wrote Bourgeois. "If nothing else there was an awareness of, and respect for, the landscape that we were coming from. The DIY punk ethos was taken very seriously and in some respects had a more lasting cultural impact in Europe than even in the US, in part because the communities are smaller and the space between them is not as vast so the networks tended to be stronger. Many of the underground touring circuits from the mid-late 80s were still around in '99 when we were there. I think also that there is more of a historical connection to the avant garde and leftist political traditions that people are still aware of."

When possible, soul-searching French people made opportunities to leave the country and spend time in these places that had captured their imagination. Sixpack drummer Maxime Charbonnier, after his band broke up, decided to leave the small provincial St. Etienne looking for a change. The previous year, the Capital City Dusters played a show in St. Etienne he had helped book. The Dusters, comprised of singer/guitarist Bourgeois, drummer Ben Azzara, and bassist Jesse Quitslund, toured extensively for much of their seven-year run between 1996 and 2003. Bourgeois worked for Dischord at the time, and with that he carried some cultural cache in addition to that which the band already had abroad, being from DC. The Dusters did an interview with Natasha Herzock for *Positive Rage #8,* which she admits greatly impacted her impression of DC:

Alex Bourgeois: Nous sommes déçus, bien sur, de ne pas jouer a Paris, mais en même…C'est ca qui est représentatif d'un pays. Nous ferons sans doute les musées plus tard. Aux States, c'est vraiment dur. Tu joues souvent sans être payé ni même nourri, et tu dois te débrouiller pour dormir. On ne te loue pas d'hôtel, et tu finis souvent par dormir par terre. On joue régulièrement dans des endroits miteux, qui puent la pisse. Nous sommes donc très contents en Europe. Nous sommes bien recus et bien mieux payés. Le contact est vraiment différent.

(Ben (batterie) nous explique que la ville est tres particuliere. Il n'y a pas de quartier noir ou latino comme dans d'autres villes des Etats-Unis.)

Ben : « Il y a la partie clean et aisee de la capitale, et une autre où l'on retrouve les gens moins aises, avec toutes les ethnies melangees. Ce n'est pas comme á New York par exemple. C'est d'ailleurs dans cette partie de la ville qu'habite la grande majorite des groupes indes. »

Alec : C'est vrai, on se connaît tous. On vit tous dans les meme quartier, et on se sencontre souvent en faisant les courses, ou chez le disquaire du coin. Tu sais, j'ai moi-meme travaillé pour Dischord, au bureau (rires)!

Alec Bourgeois: "We are disappointed, of course, not to play in Paris, but in the same way … This is what is representative of a country. We will probably do the museums later. In the States, it's really hard. You play often without being paid or even fed, and you have to manage to sleep. We do not rent a hotel, and you often end up sleeping on the floor. We play regularly in shabby places, which stink of piss. We are very happy in Europe. We are well received and much better paid. The connection is really different."

Ben (drums) explains that the city is very private. The black or Latino neighborhoods are unlike other cities in the United States.

Ben: "There is the clean and comfortable part of the capital and another where we find the less wealthy people, with a greater mix of ethnicities. It's not like New York, for example. It is in specific parts of the city that the vast majority of these bands live, however."

Alec: "It's true, we all know each other. We all live in the same neighborhood, and we often meet by shopping, or at the local record store. You know, I myself worked for Dischord, in the office!" (laughs).

When they played in St. Etienne that year, the Capital City Dusters made a great impression on Charbonnier and told him he was welcome to come visit them in DC when he had an opportunity. The following year, he decided to take them up on their offer. He arrived in Washington in Spring 2000 with travel companion Stéphane Delavacque, visual artist and author of the long-running zine *Rad Party*. The pair had no real directive other than to escape France for a while. Maxime's local habitus kept him within a steady orbit of musicians' group homes, work places, and venues (in some cases, places that overlapped all three).

Maxime Charbonnier (Sixpack) with Jesse Quitslund (Capitol City Dusters) outside of Quitslund's apartment building on 14th Street NW, Summer 2000

He spent the first few days staying with Quitslund on 14th Street NW near the Black Cat, where Jesse bartended. He then moved less than a mile northwest to Adams-Morgan, where he stayed with Bourgeois. Through Alec, he met a steady stream of people he had only heard on records and read about in zines and online. On his first day at Bourgeois' house, Maxime was looking after the house while Alec was out, and he heard the doorbell ring. He went to open the door and saw Ian MacKaye standing there, who asked for Bourgeois. All he could describe it as was "stupor," barely able to form the words in English to tell MacKaye how cool it was to meet him.

Though university would eventually drag Charbonnier back to St. Etienne by the fall of 2000, he wound up spending a total of four months in DC, learning local history, blending in, and walking up and down 14th Street. He spent much of that time span renting a room from the photographer Pat Graham in a group home at 14th and S Street, a block away from The Black Cat. He describes that house as "punk rock heaven," routinely getting to spend time with Ian Svenonius, Guy Picciotto, Brian Baker, among several others, who passed through. He also took advantage of the informal economy that circulated and had developed around the underground music scene. The Dusters' friends set him up with temporary under-the-table gigs at a French restaurant north of Dupont Circle. He also worked for a short time in an Adams-Morgan hair salon popular among many people connected to the scene, further reflecting the profound social dynamic to cultural circulation, connecting with people and their social practices.

Charbonnier, for someone in a marginal geographic position within France's punk scene, came to DC and almost immediately became a part of the mythologized scene he had grown up hearing about and absorbing through recordings and zines. His affection for DC and his then-alienation from St. Etienne went hand in hand. Maxime still acknowledges, however, that foreign experiences of place in the US are heavily predicated upon personal and empirical connections.

"I loved the city," he wrote. "I loved every minutes [sic] of my stay. The city is beautiful, lots of things to do if you're into art [and] culture. Cruising the city with my bike were priceless moments. I loved the smell (which can be challenging), the weather …everything. But, if not introduced that well as I was, it may [have been] a little hard to find the good spots and meet the right people."

Charbonnier's experiences in DC were obviously singularly fortunate, as he entered the scene as a fan trying not to seem too much like one, yet within weeks sharing a group home with some of his heroes. This speaks to a greater social aesthetic and approachability that most music scenes in the US simply did not have. This also speaks to a valuable lesson about the circulation of musical culture, in that music operates as embedded within social and spatial lives, not apart from them. The *habitus* of those in the DC underground scene transcended music, as the places of their daily lives became part of one fan's permanent impression of the landscape. When Charbonnier first met the Capitol City Dusters, all he knew about the Dischord House came from photographs in Minor Threat liner notes and small-press photo books. The next year, he and Steph Rad Party were hanging out with Alec Bourgeois, Cynthia Connolly, and other luminaries at the label office.

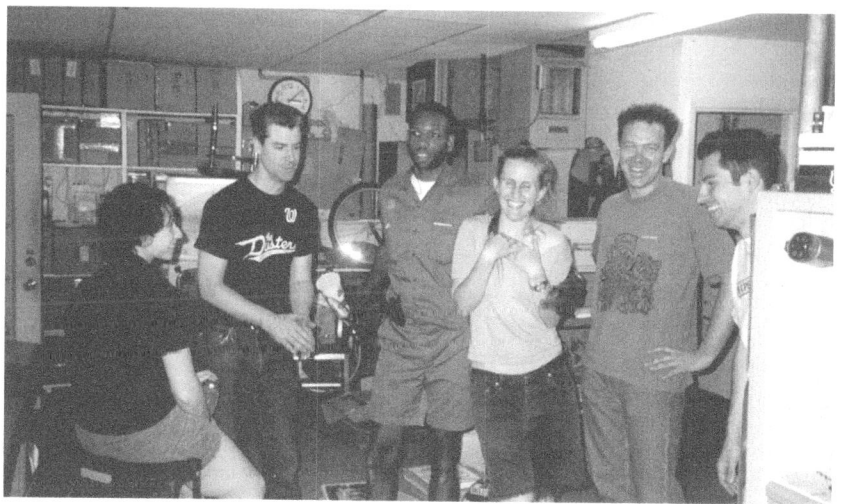

The Dischord Records office receives a pair of visitors from France, Summer 2000. (L-R) Melissa Quigley, Alec Bourgeois, unidentified UPS driver, and Cynthia Connolly with Steph Rad Party (Stephane Delevacque) and Maxime Charbonnier (St. Etienne). (Image courtesy of Charbonnier)

Maxime's surreal luck even transcended the local and roped in a globally recognized pop icon on his second day in town. He and Steph went to visit the Smithsonian American History Museum, where he remembers finding the leather jacket worn by Henry Winkler playing Arthur Fonzarelli (aka "The Fonz"). *Happy Days*, being an internationally popular show, had impacted much of the European imaginary of 1950s Americana. An hour later, he was walking down the street on the way back to Quitslund's house and he spotted someone who looked familiar. It was Winkler, smoking a cigar outside of his hotel. He was in town to appear in a staging of Neil Simon's "The Dinner Party." Maxime approached Winkler, who was very enthusiastic to learn this young man was from France, where he spent a lot of time off over the years.

The layers of symbolism embedded in this encounter merely begin with how Winkler, a Jewish actor best known for playing an Italian American character, represented the synergy of cultures that incubated the New York punk scene (see Beeber 2014). On a deeper level, Winkler and the rest of the *Happy Days* team had a profound impact on the French imaginary of *L'Amerique Profonde*, and it all came home in one chance encounter. Though Maxime still laughs about the serendipity of that encounter, it still dominantly shapes his personal sense of place of DC more than any experiences visiting DC's overcrowded sites of key tourist gaze.

The DC visits of French punk musicians and fans like Maxime Charbonnier, Hugo Maimone, and Maïe Perraud were of varying significance, motivation, and quality. Every agent of cultural circulation has their individual set of experiences that inform their sense of place. Some, like Perraud, were visiting for mainstream reasons and unimpressed with a city they did not know had such a vibrant underground scene. Others like Maimone were in DC for reasons unrelated to punk, but got to dig into the punk landscape in parts. For others, like Charbonnier and Delevacque, the District brought so many American fantasies to life that had it not been for photographic evidence, their friends back home would not have believed them. Just as the push-and-pull factors of tourism lie on a spectrum, so do the individual relationships between visitors and their destination. Whether the tourists visit the city motivated by its musical sites, or whether their visit preceded their education about the underground scene, the circulation of urban ethos relies on complex sets of representations and individual experiences.

Bibliography

Beeber, S. L. (2014). Juidos 'n' Decaf Italians: Irony, Blasphemy, and Jewish Shtick. In B. Lashua, K. Spracklen, & S. Wagg (Eds.), *Sounds and the city: Popular music, place and globalization* (pp. 76–91). London: Palgrave Macmillan.

Dunn, K. (2016a). Email correspondence, 16 June.

Dunn, K. (2016b). *Global punk: Resistance and rebellion in everyday life.* New York: Bloomsbury.

Gotham, K. F. (2005). Tourism gentrification: The case of New Orleans' vieux carre (French quarter). *Urban Studies, 42*(7), 1099–1121.

Knupp, R. E. (1981). A time for every purpose under heaven: Rhetorical dimensions of protest music. *Southern Speech Communication Journal, 46*(4), 377–389. https://doi.org/10.1080/10417948109372503.

8

"We Were Fucking Tourists, in the End": Punk, Tourism, and Gentrification

Thrashington DC in Washington, DC, July 2009: (L-R) Fabrice Le Roux (Fast Fab), Timothée Priol, Thomas Laguerre (Goose), Régis Rollant, Lionel Cadiou. (Photo by Yann Bernard. Image courtesy of the band)

© The Author(s) 2019
T. Sonnichsen, *Capitals of Punk*, https://doi.org/10.1007/978-981-13-5968-2_8

Punks and Tourism in the Time of Gentrification

Agents of cultural circulation, including members of Fugazi and Shudder to Think going one direction or members of Heimat-Los and Krüll going the other, are engaging in different shades of tourism. Their objectives and directives all varied, and the only common thread may have been music, but they were all what Tim Edensor (2009) refers to tourists as "semioticians." These engagements with urban landscape within the framework of punk all suggest how vital tourism is for the construction and enactment of cultural identity.

The transatlantic circulation of the underground between two capital cities is a great platform through which to understand the underlying nature of tourism, as well. Both geography and tourism studies have an ample index of work that studies the intersection of the two with music. Australian geographers John Connell and Chris Gibson (2004) have published extensively on how music drives the relationship between tourism and sense of place, how "through a recourse to nostalgia and claims to authenticity, music may also reveal ideas about the nature of tourism" (p. 3). Tourism, considering how it went largely ignored in both geography and musicology until this century, also proposes myriad challenges. Because both the practices and practitioners of music tourism are so varied, to conduct any kind of straightforward quantitative analysis is impossible. Relatively little concrete data exists outside of standardized, centralized markets for music consumption. For example, ticket sales figures and demographics for a major music festival like Glastonbury, Coachella, or Big Day Out, while often proprietary, exist and can be manipulated for market research purposes. Once anybody starts conducting research on smaller venues and DIY shows, especially from the twentieth century, documentation can often be full of holes. Oral histories, are often unreliable, but often the only extant data source. Though digital ticket sales and social media have made punk tourism more trackable in the twenty-first century, there are still gaps between qualitative and qualitative measures on the historic status of an alternative destination like the Dischord House or the Black

Cat. Music tourism remains an amorphous subject. In 2007, Gibson and Connell offered an attempt at a definition: "A range of practices where sites of music production and expression (whether in past or present 'scenes') become the points of attraction for tourists, and may also become central to strategies employed by the local state, tourist promotion boards and companies to market musical heritage and a musical environment" (p. 167).

Ultimately, music has become a motivating factor for the progression of tourism research. For many years, a disproportionate amount of data collected and analyzed on tourist activities performed and observed on-site ignored many of the geographic, sense-of-place factors that motivated visitors to be there. As Graham Dann lamented 40 years ago, "few investigations begin with the question, 'What makes tourists travel?'" (Dann 1977, 185).

The act of touring, for example, is rarely considered in the conversation about music tourism. Tourism is frequently understood as a series of acts of consumption (sights, sounds, food/drink, and experiences), rather than as an active creation. Fugazi toured Europe extensively, motivated by the desire to bring live music to as many people and places as possible. Their motivation was ostensibly financial, but it was genuinely driven by their belief in music circulation and geographic and cultural expansion and exploration. A big disruption occurs because touring artists spend, at most, 20% of their waking time actually onstage. Outside of promotional activities (signings, interviews, and recording), an artist's daily schedule does leave downtime where the artists have the opportunity to engage in traditional tourist practices. Punk networks, however, provided ample opportunities for those among the underground exchange to be nontraditional tourists, or at least engage with less traditional practices as both artists on tour and fans visiting music-related sites. All tourism resides on a spectrum in between "Fordist" practices (prepackaged, standardized, guided, manicured, and often whitewashed) and "post-Fordist" practices (unmitigated, unsupervised, or immersive; see Torres 2002). In turn, the interdisciplinary field of tourism studies has increased focus on specialized and otherwise niche tourism. In simple terms, this spectrum of touristic practice illustrates, on multiple levels, the dichotomy between "the rough" and "the clean." Tourism frequently draws from both conventional, "clean" activities like staying in a hotel and visiting museums versus "dirty" activities like sleeping on friend's couches, visiting local dives, or engaging in potentially illegal activities.

The "Clean" and the "Dirty" in Tourism

Washington, DC, and Paris both provide opportunities for visitors to engage in both "clean" and "dirty" tourism. Paris, particularly the Pigalle neighborhood in the shadow of Montmartre, grew into a mecca of cabaret-laced hedonism at the end of the nineteenth century. Legendary dance halls and theaters included Le Moulin Rouge, Les Folies Bergeres, Le Chat Noir, and Le Cirque Medrano. Between 1892 and 1900, the number of brothels in Pigalle ballooned from 59 to over 100 with Moulin Rouge at the core. Though Henri de Toulouse-Lautrec's paintings of the establishment have remained in the public eye, Baz Luhrmann's 2001 film based on the Moulin Rouge's fin-de-siècle mythology did wonders to increase visibility of the legendary cabaret in contemporary popular culture. Though the provocative can-can dance originated there, the original windmill atop the building burned down toward the beginning of World War I. Today, tourists herd in and out of a simulacrum in record numbers, subjected to excessive costs for nearly every activity within the walls. Meanwhile, other strip clubs and flophouses line the Boulevard de Clichy, providing libidinous attractions for stag parties while benefitting from the district's accessible presentation of a quintessentially "dirty" tourist activity.

Paris' institutionalization of its red-light district is a similar paradox found in other European cities, especially Amsterdam and Hamburg. Prior to the end of World War II, Americans best knew all three cities through stories from troops and naval officers who went ashore in these towns and found a wide range of vices ready and waiting. For many of the young men who had grown up with little exposure to American inner cities, this was a pivotal point in their lives. As the song rhetorically asked, "How you gonna keep 'em down on the farm (after they've seen Paree)?" so did the newfound sexual freedom afforded these young visitors. Hamburg became Germany's epicenter of hedonism for these reasons.

The Reeperbahn, home to many seedy nightclubs where the young Beatles would hone their musical chops in the early 1960s, became such an institution for visitors that local preservationists even became invested in it. Facing the current wave of gentrification and private sector investment among cities in Western Europe, Fremaux and Fremaux (2013) noted how "the St. Pauli Preservation Society...laments the decline of the Reeperbahn's hedonistic heritage" (p. 304).

Though somewhat contradictory that sex and drugs would become such a driver for international tourism, cities still found manners through which to present these institutions less problematically. Though Washington, DC, lost one red-light theater district on 9th Street near downtown around World War II and the other on 14th Street NW has become heavily gentrified, other US cities like New Orleans and Las Vegas have successfully marketed themselves as hedonistic party towns in the vein of Amsterdam. This process belies a wide spate of serious quotidian issues faced by the city's residents in the name of securing greater investment from curious outsiders. Kevin Fox Gotham (2005) wrote that the New Orleans French Quarter successfully "impart[ed] an unthreatening and unproblematic representation using themes such as romanticism, nostalgia and other flattering images" (p. 1744). In doing so, these cities have played down prevalent problems like drug addiction, homelessness, and prostitution systematically.

DC and Paris both lean heavily on tourism industries which necessitate safe, unproblematic versions of their landscapes. Itinerant visitors are normally aware that their experience is often unrepresentative of "the authentic" experiences of lifers in those towns, which are, normally, less than desirable for a carefree holiday. The eighth-grade trip to DC is an American tradition for anyone within a days' drive of the Capital (i.e. most anything east of the Mississippi River). Paris is a similar magnet for French students coming of age and learning about cultural and civic life. For people who live in these cities, the tourist icons have little to contribute to their daily life; they are just "there." When I lived in DC, I only visited the Smithsonian Museums (save for the Portrait Gallery, which was convenient to my lunch breaks) when I had friends and family visiting. Olivier Firminhac, the owner of Crapoulet Records in Marseille, proudly told me that in almost 30 years of living in Paris, he never once went to the Eiffel Tower.

Tourism is generally mutually exclusive with the concerns of the quotidian, because most travel experiences are based off of imagining and experiencing new, unfamiliar landscapes. Punk tourism, as a significant splinter of musical tourism, suggests that this dichotomy could actually be less rigid than travel writers and tourism theorists have suggested. I noticed this in the reflections of several collaborators who had never been to DC, or even the US. Orléans native Noémie Ventura, a photographer, singer, and longtime Dischord collaborator living in England, has never made the journey stateside. Still, her impressions of Washington, DC, are heavily influenced by stories passed through the punk scene and representations in Fugazi's lyrics.

"I think that my perception of it is very much based on all the stories and interviews I have read from all these very interesting people," Noémie wrote me recently. "I think that there used to be a very active punk [and] hardcore scene in the 80's and 90's, and it seems that there are still quite a lot of free thinkers and independent spirits there. I imagine that it may be a natural consequence of having the US government based there. This city must be very monitored, so as a reaction there is a strong indie scene. That's my theory, anyway."

Two vital voices in this conversation about punk-bred perspectives on DC are Natasha Herzock and Mathieu Gélézeau. They shared guitar and vocal duties in the post-hardcore band Kimmo, sounding so much like home to me when I saw them play on my first trip to Paris on that night in 2010. They also worked together for years on the zine *Positive Rage*, whose name referred to the DC activist collective Positive Force, bringing attention to their favorite underground bands on both sides of the Atlantic. Mathieu's story about Dischord Records in a 1998 edition of *Positive Rage* was, to my knowledge, the first relatively comprehensive French-language history of the label. He also spent years running a record distro called No Reason after the eponymous Minor Threat song. Since putting Kimmo on hiatus, Mathieu and Natasha have occasionally performed electronic music as Computerstaat. Despite a decades-long devotion to a roster of artists from the DC underground scene, neither has been to visit the Capitol. Gélézeau is not particularly interested.

"In France, [our perspective on DC is] strange because New York – it's more famous," he said. "Washington, DC is the capital, but it's not very interesting for [many French people]... I don't want to go to DC because I [wouldn't] like to go to just say 'Hello, I'm a fan.' If I [collaborated on] something with [artists] there, then I [would] go there. But not just to, you know, 'click click clack,'" he added while mimicking a tourist snapping photos, "I don't want that. If I go there, I want to do something [artistic] there."

Natasha is more open about the prospect of visiting DC, having only been to Los Angeles on holiday and to Texas for a work conference years ago. Similar to Mathieu, interviews she has done with Dischord bands have influenced her imaginary of Washington.

"I can imagine things [about DC], but I don't know if it's real," she mentioned. "I can imagine things with music and the people I've met. It was a really interesting interview with the [Capital City] Dusters, because they told us about geography and how people mixed. Washington sounds to me like an interesting town with what the Dusters told us."

There remains little room for debate about the impact that punk music has had on approaches to tourism. Granted, music tourism is still often at the crux of the discourse over the "clean" and the "dirty" presentations of tourism. However, visiting a site because your favorite band was photographed in front of it or because they sang a song about it or merely mentioned it may be among the purest forms of tourism, or at least the most high-involvement. Music's ability to influence emotional attachments to unfamiliar sites in unfamiliar cities is paramount. It demonstrates the power of all media representations in informing narratives of place, and how music frequently helps add nuance to those impressions. It is at that cultural crossroads where counternarratives develop and separate themselves from mainstream presentations and interpretations of these cities.

A place like the White House, one of the most photographed buildings in the US, stands as a prime example. Images of the structure, especially those which carry near-universal recognition in the West, can mean radically different things to different people. Political leanings and shared memories are just two of many factors which influence how one may interpret the actual site or even a representation of the White House via social media, film, or television. Ray Hudson (2006) wrote how National icons like these prove how "places are contested and continually in the process of becoming, rather than essentialized and fixed, open and porous to a variety of flows in and out" (p. 627).

From a conversation I had in Marseille in 2015 with Olivier Firminhac and Claire Samant

(Crapoulet Records):

Claire: We've never been to DC before, but we saw 'Independence Day.'
Olivier: You can imagine a city full of people who are working for government, I think it may be the case, and not the fanciest city because aliens tend to go to San Francisco or Los Angeles – and for one movie they go to Washington, so I don't think it's a fancy city. But I think it's a quiet city, too, except for what happened with hardcore, but I think all the rest is pretty calm. And quiet.
Claire: I guess it's like if you talk to people from abroad, about Australia, we would say Sydney is the capital city but it's really Canberra, but we forget because we don't really care [laughs]. I think people sometimes forget DC is the Capitol City.
Me: Really?
Claire: People think it's New York. It's the biggest city, people think – you know particularly for France, Paris is the capital, and is the biggest city, and is the most cultural – and Washington look some small city of countryside, something with old ladies, it doesn't look that cool, actually. When we see it on TV shows, it's always something with politics, or White House destruction…
Me: You really like Independence Day, don't you?
Olivier: You notice when a movie is in New York, the monster destroys New York.
Claire: King Kong, too!
Olivier: Or like Superman, what he does fly over New York… when a movie is in Washington, it destroys one house. It is one big house, and it's over, and you don't have any images of the rest of the city.
Claire: For me, it just looks really quiet, just a bit boring, if you compare it to New York.

Many French musicians in this generation of punk have gotten tangled in some surprising intersection of narratives at places like these. In August 2009, Thrashington DC headed to the US for their only American tour as a band. The morning after playing a basement show in the District, they decided that no tour for a band so referentially named would be complete without a group photo in front of one of America's most iconic buildings.

"We met a guy just in front of the White House, he was like 40 or 45," recalled Fab Le Roux. "I was wearing a Black Flag T-shirt, and he stopped me and said, 'Black Flag, cool band! I used to play in Millions of Dead Cops.' I don't know if it was true or not, but we just thought it was funny, talking to a guy about Black Flag in front of the White House. We wanted to do the photo in front of the White House. We were fucking tourists, in the end."

The White House bears a strong resemblance to le Chateau de Rastignac outside the provincial town of Périgeaux, which Le Roux introduced to me. Apparently, historians continue debating whether the Chateau directly influenced the White House or the other way around, depending on their respective dates of completion and state visits. The White House (1802), completed in time for Jefferson's term as president, was completed 15 years before the Chateau (1817), but the commonly accepted theory is that Thomas Jefferson saw the designs for the Chateau on a state visit in 1789. Ironically, the Chateau has become a popular tourist destination for Americans and other foreign nationals, a bizarre remnant of the transatlantic circulation of neoclassical architecture.

Gentrification and Underground Circulation

Punk presents itself as one of the best investigative objects for addressing the urban space from a musical point of view. Born in the United States and England, around the cities of New York and London, and in the midst of deep political, economic and social crises of the 1970s, it is connected to a political, economical [sic] and social system in deep and disorienting transformation. By keeping distance of a utopian revolutionary positioning, punk sought in the reality of urban chaos, the basis for its constitution. (Dos Santos 2015, 136)

It would be impossible and irresponsible to omit discussion on gentrification as an omnipresent dimension within the discourse about Washington, DC, and Paris. Arguably, the political, social, and physical conditions of the cities were what made punk and hardcore scenes grow and circulate prior to the heavy onset of neoliberal redevelopment toward the end of the century. Today's accelerated market-driven development, which has seen money, development, and related human capital pour into these cities, the spaces and places where underground music thrive have been deeply compromised. Simone Krüger (2014) noted how attempts "to offset these negative consequences of globalization are … conceptualized under the theme of 'creative cities'" (p. 136). Those who propel creative cities have been immortalized in academia and policy as "the creative class," most famously by geographer Richard Florida (2002). In these cases, circulation has become an inextricable component of these changes in urban landscape, both positively and negatively. The ways that cities like DC and Paris have reinvented themselves have depended largely upon how city dwellers reinforce and interpret these circulations of capital and branding as modernizing forces. In order for cities to remain viable in a rapidly expanding global economy, they must embrace outside influences and investments, however begrudgingly. Perhaps the only consistent facet of urban life in the twenty-first century is change.

Both DC and Paris are among the most expensive places in which to live in their respective countries. Due to zoning laws, space in either was at a premium even before gentrification. The DC Metro region expanded quickly in the wake of World War II; the region tripled in population between 1940 and 1970. However, this growth was predominantly suburban, as Maryland and Virginia suburbs grew about 700% while the District only grew 14% over the same time frame. Both cities profoundly restrict skyscrapers within their respective district borders. Outside, however, thoroughly modern lite skyscrapers dominate peripheral skylines.

The capitalist orientations for development, however, vary slightly. Traditionally, in Western Europe, the wealthiest landowners have held onto their downtown property, and many middle and underclass were never able to live in the inner city. Therefore, as the wealthy elite in Paris "have always favored central neighborhoods for their residence, … the idea of a movement 'back to the city' seems inadequate for social categories that never left it" (Préteceille 2007, 12). Until somewhat recently, Paris has had some exceptions, including Belleville and other locations on the eastern end. In his landmark work on the revanchist city, geographer Neil Smith (1996) echoed a common assertion that Paris presents a French contrast to the traditional Anglo-Saxon model that treated the city like a cell, emanating on varying class-based ring layers from the Central Business District. He elaborated on the contrasts, which included Europe's stricter urban building codes, more laissez-faire treatment of urban land and housing markets in the US, and differing histories of racial segregation.

It follows, then that gentrification in Paris should be considered different from gentrification in DC, even if both are motivated by greed and traditional treatment of urban areas as "growth machines" (see Molotch 1976). Development and gentrification of Paris is also couched with discourse on *embourgeoisement*, in which the city systematically redefined itself at the behest of old money and the middle class in pursuit of that lifestyle. The majority of French research on gentrification, embourgeoisement, and the intersection of the two has called to attention how Anglo-American cities tend to favor a supply-side approach to redevelopment, where cities in continental Europe have relied on more demand-side.

Especially in the 1980s, when various "islands of renewal in seas of decay" (see Smith 1996, 95) throughout northern and eastern Paris had disintegrated into warzones for the younger generation, the heavily militarized urban palaces that remained from the turn of the prior century presented a strange dynamic for middle-class and working-class punks.

"Paris has too many people, flats are very expensive and small, and it's too crowded," said Herzock. "I used to live in Paris when I was a student. I was listening to punk rock music, and it was not the temper of the area [in the 12th Arrondissement]. It [was] very bourgeois. And I was punk. I like classical music, but..." she trailed off, laughing.

It was not until the 1980s that many visible forms of gentrification accelerated in Paris. Fittingly, as in receiving, adapting, and interpreting American popular music, gentrification was another method through which it lagged behind other Western cities. Exacerbated class divisions that resulted were among the obstacles that stunted the growth of the underground punk and hardcore scene. Philippe Roizès, who grew up in a lower middle-class family in the western suburb Colombes, was often unable to stay in the city for punk shows late at night. If shows took place on weeknights, the last Metro train that could get him back to Colombes left well before midnight. Hardcore matinees at places like La Cithéa were more accessible, yet still skinhead violence made eastern Paris high-risk for young people at any time of day. Roizès was fortunate, however, to attend high school in Paris close to New Rose, the city's pre-eminent punk boutique. The shop, also often swarming with skinheads, became a valuable point of cultural exchange for Philippe and his new group of friends from the city. The suburbs, though they had plenty of punk fans, offered little in the way of cultural hubs or meeting places. Still, for some, the suburbs became shorthand for underclass cohesion and rebellion.

"I think there was a difference between the people living in the suburbs and the people in Paris," said Stéphane Delevacque. "The people in Paris were richer, and had access to more culture. Most of the people I met at the time [who liked hardcore], we met in Paris, but we were all living in different parts of the suburbs. I would have loved to be able to live in the city, because living in the suburbs, it was always a nightmare to come home after the gigs and sometimes miss the train and have to sleep on a bench and catch the first train back in the morning. Most people who were into hardcore were the outcasts, they were not the ones who went into school and work. Being out [in the suburbs], it gave us a realistic vision of what we could do, trying to build something with a different approach than money and things."

The suburban punks' use of Café Verdun as a meeting place every Saturday in the mid-to-late 1980s demonstrated how they needed the city in order to foster their community, however. Locations like Verdun in Paris, like the Dischord House in Arlington, Virginia, were both epi-centers for the expression, networking, and ultimately circulation of punk between the two cities. The ratty Cafe Verdun closed down years ago, and much like the rest of the neighborhood around Gare de L'est., has reinvented itself as a finely manicured signpost of gentrification. Pre-gentrification spots like Verdun are proof that "when 'reading' the cultural landscape, it is easy to forget the impact of that which has been rejected" (Burman 2010, 104).

Punk music, like the rock 'n' roll tradition that preceded it, often set itself against a "backdrop of urban decay and isolation, [and] to better understand this relationship, we need to look at the nature of the city" (Keeffe 2010, 145). The landscape of late-1970s DC that created the subculture which remains so unchanging and mythologized in the minds of punk fans worldwide has given way to almost cartoonish degrees of gentrification, replete with spiraling real estate values. Between 2000 and 2012, home prices in Mount Pleasant and Adams Morgan, both crucial neighborhoods for the proliferation of DC's underground culture, rose from a median price of $200,000 to $530,000. Over the same period, median home prices in adjacent neighborhoods like Petworth and Brightwood Park went from $186,000 to $455,000. Rents throughout the District responded, pushing out many longtime residents and preventing new, potentially creative ones from moving in. The house in which Minor Threat played their first show in 1980 at 1929 Calvert Street NW would sell, more than 33 years later, for a price that nobody at the time would have believed was possible.

Legendary performance spaces for this unique punk identity, deeply rooted in place, have mostly vanished as the District area has proven that, as Ian MacKaye sang in Fugazi's "Cashout" (2001) "what development wants, development gets." Even the 9:30 Club has long since left its original downtown location at 930 F Street NW and moved into an old radio station building on V Street close to Howard University. Symbolically, the nearby Howard Theater, one of the first integrated music venues during the city's pre–civil rights era, fell into disrepair over this time period. By 2013, a well intentioned local coalition would save and restore the theater as a testament to the city's reverence for its black musical tradition. These publicly funded motions to "protect" DC's increasingly recognized musical heritage unwittingly reinforce an uncomfortable reality about how many local black institutions, including the Howard Theater, Ben's Chili Bowl, and Bohemian Caverns have been subsumed into a white, middle-class cultural exchange system (see Regis and Walton 2008). DC's most mythologized underground musical movements, namely, harDCore and Go-Go, have been rediscovered on a civic level as a fulcrum for symbolic urban landscape preservation. The DC Public Library, which has embraced and presented these histories in varying degrees of museafication, has accomplished much in legitimizing these forms as cultural significant. However, much of this has happened as a direct result of, and reaction to, gentrification, drawing upon local histories in such a way that fulfills contemporary demand.

As was common of North American cities after World War II, DC's suburban development and freeway construction drew the predominantly white middle and upper class out to the periphery of the city. By the end of the 1950s, most cities had taken a substantial demographic, architectural, and administrative hit; Washington stood out as one of the east coast's most egregious case studies. When they sought to reinvent themselves in this new neoliberal climate, American cities under the so-called "City Beautiful" movement actually objectified icons of European urban design. As early as 1958, however, Jane Jacobs was issuing warnings that it would be a mistake for American cities to look to the boulevards of Paris for a model of new urbanism.

Urban renewal in the US, like the European version of it that soon thereafter hit Paris (as much as World War II had aided in large-scale destruction), was fraught with these types of contradictions. In the 1960s and 1970s, the major trends in urban redevelopment and late modernist design belied how that time was "a watershed in the institutionalization of urban fear" (Zukin 2013, 356). For the past three decades, regardless of crime statistics and on-the-street lived experiences of the cities, redevelopment has shaken older modernist ideals. Under new urbanism or neo-traditionalism, planners have made blatant attempts to turn back the clock, however impossible, to the pre-automobile era in an effort to undo generations of infrastructural change. Though some American colonial cities like Boston and St. Augustine have translated their preindustrial geographies into tourist revenue, most of the US, DC included, came into existence during the industrial revolution. Much preindustrial nostalgia prevalent in American cities, seen in attempts to centralize population and reinstitute public transit, takes as much a cue from quintessentially "European" cities as it does pre-"White Flight" American ones. The respective peripheries of Paris and DC, though both more Metro-accessible now than 30 years ago, did not enjoy the windfalls. Though the US Department of Housing, looming over the inequality mollifying throughout DC, issued a report in 1979 devoted to mini-mizing displacement of vulnerable individuals, it is still estimated that in 1985 gentrification was displacing over two million people per year in the US.

To those fortunate to have moved *into* metropolitan DC during the city's postmillennial era of increasing gentrification (this writer included, in 2005), the idealized pasts of harDCore and Go-Go have provided a forum for disalienation. In Paris, a similarly bleak outlook permeates local music. Frustration, one of Paris' most popular synth-punk bands, eulogized the old Paris in their song "Dying City." Such a sentiment about Paris' constricting noise and zoning laws was nothing new. In late May 1986, Manu Casana wrote to Dischord, angry over Jacques Chirac's crackdown on smaller clubs; cops had pulled the plug on a Sherwood gig eight songs in due to a fire code violation. Twenty years later, as a response to gentrification, activists would place flyers on doors of priced-out bars and restaurants that read "FERMÉ POUR CAUSE DE VILLE MORTE. Merci de vous adresser à la capitale d'à côté..." ("closed because of the dead city, please address yourself to the capital next door...") (Straw 2010). The underground music scene coalesced in response to the domineering mediocrity of the mainstream, and so was this process mir-rored in the palimpsest life of the city. Also noteworthy was how many spaces of artistic creation were being supplanted with spaces of food and drink consumption.

Anti-gentrification movements, largely steered by artist-activists, rise and gather momentum in the face of development. In this way, cities like DC and Paris, often lamented as different from the "old" iterations of the towns by scene veterans, provide transforming platforms for these cul-tures to circulate. Ironically yet predictably, the marginal voices and scenes get squeezed as "urban centers become spaces of culture and the arts, spectacle and play, conspicuous consumption and the accumulation of collective symbolic capital" (Markley and Sharma 2016, 389). The ways in which that circulation has changed tends to reflect how the cities themselves have changed. Although punk and the rock 'n' roll foundation that preceded it both sanctified the "ugliness" of the city, could either ever thrive in "the city beautiful"?

Bibliography

Burman, J. (2010). Absence, "removal," and the everyday life in the Diasporic City: Anti-detention/deportation activism in Montreal. In A. Boutros & W. Straw (Eds.), *Circulation and the city: Essays on urban culture* (pp. 99–117). Montreal: McGill-Queen's University Press.

Connell, J., & Gibson, C. (2004). Vicarious journeys: Travels in music. *Tourism Geographies, 6*(1), 2–25.

Dann, G. M. (1977). Anomie, ego-enhancement and tourism. *Annals of Tourism Research, 4*(4), 184–194.

Dos Santos, D. G. (2015). Between drums and drones: The Urban experience in Sao Paulo's punk music. In P. Guerra & T. Moreira (Eds.), *Keep it simple, make it fast: An approach to underground music scenes* (Vol. 1, pp. 135–145). Porto: Universidade do Porto – Faculdade de Letras.

Edensor, T. (2009). Tourism. In R. Kitchin & N. Thrift (Eds.), *International encyclopedia of human geography* (Vol. 11, pp. 301–312). Oxford: Elsevier.

Florida, R. (2002). *The rise of the creative class*. New York: Basic Books.

Fremaux, S., & Fremaux, M. (2013). Remembering the Beatles' legacy in Hamburg's problematic tourism strategy. *Journal of Heritage Tourism, 8*(4), 303–319.

Gotham, K. F. (2005). Tourism gentrification: The case of New Orleans' vieux carre (French quarter). *Urban Studies, 42*(7), 1099–1121.

Hudson, R. (2006). Regions and place: Music, identity and place. *Progress in Human Geography, 30*(5), 626–640.

Keeffe, G. (2010). Compost city: Underground music, collapsoscapes and Urban regeneration. *Popular Music History, 4*(2), 145–159.

Krüger, S. (2014). Branding the city: Music tourism and the European Capital of culture event. In S. Kruger & R. Trandafoiu (Eds.), *The globalization of musics in transit: Music migration and tourism* (pp. 135–159). London: Routledge.

MacKaye, I., Picciotto, G., Canty, B., & Lally, J. (2001). Cashout. Prod. Zientara, Don and Fugazi. *The Argument*. Washington, DC: Dischord Records.

Markley, S., & Sharma, M. (2016). Keeping Knoxville scruffy?: Urban entrepreneurialism, creativity, and gentrification down the urban hierarchy. *Southeastern Geographer, 56*(4), 384–408.

Molotch, H. (1976). The city as a growth machine: Toward a political economy of place. *American Journal of Sociology, 82*, 309–332.

Préteceille, E. (2007). Is gentrification a useful paradigm to analyse social changes in the Paris metropolis? *Environment and Planning A, 39*(1), 10–31.

Regis, H. A., & Walton, S. (2008). Producing the folk at the New Orleans Jazz and heritage festival. *Journal of American Folklore, 121*(482), 400–440.

Smith, N. (1996). *The new urban frontier: Gentrification and the revanchist city.* London: Routledge.

Straw, W. (2010). *Cities of the night, cultures of the night.* Talk given at the 'My City's Still Breathing: A symposium exploring the arts, artists and the city' conference. Winnipeg, MB.

Torres, R. (2002). Cancun's tourism development from a Fordist spectrum of analysis. *Tourist Studies, 2*(1), 87–116.

Zukin, S. (2013). Whose culture? Whose city? In J. Lin & C. Mele (Eds.), *The urban sociology reader* (2nd ed., pp. 349–357). London: Routledge.

9

Ian MacKaye Is Alive and Well and Living in DC: Concluding Thoughts

A few months after my fieldwork in France ended, I sent an email to thank and follow up with Philippe Roizès. His response:

> *Because of your interviews, Roman [Jaskowski] and I have been deeply back into the DC scene. And we started to work together on a project of a tribute show to the DC scene. The idea is a backing band (Roman will be on bass) playing DC cover songs for the whole show with guest singers of the French scene (including Manu [Casana], Norbert [Mension], Stéphane [Rad Party] and I). Songs and guests will be from the 80's to the nowadays scene. So there will be singers from 20 to 55 years old performing DC material. All because of you!*
> *Best,*
> *Philippe*

This book began in earnest as doctoral research about how punk affected French perspectives on Washington, DC, and how that reflected back on the circulation between that city and Paris. I had no intention to facilitate musical practice, but whether or not Philippe and Romans' DC tribute was a success, it inevitably did. Though American Hardcore has inspired so much more archival work and mythologizing (including an eponymous 2006 documentary) than French hardcore, both were equally rooted in honesty in societies the progenitors viewed as anything but. As

© The Author(s) 2019
T. Sonnichsen, *Capitals of Punk*, https://doi.org/10.1007/978-981-13-5968-2_9

Kromozom 4 yelled, "Tout Faux." In DC from 1979 to 1983 and in Paris from 1984 to 1988 (timespans subjective, of course), hardcore was not a career opportunity, but both a statement and something that people just *did*. Though conduits for distribution have widened dramatically, little has changed in that regard. Players, organizers, zinesters, and friends got older. Everyone I spoke to looked back at, or imagined, that era fondly, but nobody dwelled on it. So, at least for my French collaborators, the opportunity to sit down and tell their stories brought out emotions and epiphanies. Though some ethnographers try to minimize it, human research has an impact on the informants that is well outside the control of the researcher.

As I discovered over the course of this book, a substantial part of ethnographic and geographic research is about "rediscovering" and "reassembling" subjects. As much literature on ethnography and oral histories has alluded, individual mobilities have led people away from those geographies through which these subcultures circulated. I cannot expect all of my informants to remember every detail, and as helpful as it may be in authoritatively retelling their stories, nor should I necessarily want them to. One of the gifts of viewing music as a constant circulation is the understanding how this cultural exchange still matters today and is still prone to interpretation and reinvention.

As the field of music geography expands, it will continue to embrace and encompass other fields of study and areas of inquiry, both academic and popular. In fact, studies that approach punk from an intellectual (perhaps, as appropriately, classically French) perspective have continuously disintegrated barriers between academia and pop culture. Like how the mainstream and underground need one another in order to thrive or even exist, so are pop culture and theory bound together.

The cornerstone of a more effective understanding of musical diffusion is conceptualizing music in light of the cities it represents. Music is perpetually circulating and changing in the manner which cities perpetually cycle in capital, resources, and, above all, people. This model, like all models, has contradictions; trying to capture particular moments in any circulation is nearly impossible, akin to taking a comprehensive qualitative "snapshot" of a city block. Geographic realities are highly subjective, and lived experiences around both urban landscapes and music scenes are individualized. More research should acknowledge this.

Today, the punk scenes in DC and Paris are both producing great music and inspiring politics, but the "DC sound" and the "Paris sound" have both changed dramatically. This should surprise no one; DC and Paris have both changed dramatically. Both towns will always have loud-fast-rules bands playing in the style of Minor Threat or Sherwood Pogo, but this is not necessarily because they are trying to relive the past. This is simply because those bands were inspired and driven to high levels of innovation and inspiration by local geographies. This is also because both scenes' histories have been made accessible through a growing cultural circulation. As the internet has sped up, so has the movement of music and ideas. Accordingly, this book needed to focus on the more traceable twentieth century, with the benefit of decades of hindsight from its early participants. As DC has especially demonstrated, cultural stewards and music fans at large have much to gain from investment in oral histories and the archive. Musicologists and historians have been engaging with this for some time, and I'm grateful to see geographers, media theorists, and those in other fields following suit.

Ultimately, the only urban experience anywhere close to universal in the twenty-first century has been change. Despite understandable "dying city" lamentations, the processes of redevelopment and growth, no matter how sustainable, are driving urbanites to find methods of self-actualization. "The capital of the nineteenth century" is finding creative ways to reinvent itself in the face of neoliberalism and unprecedented transformation in the twenty-first century. One could say the same for DC; privatization, inequalities, and spaces of struggle are certainly nothing new in either town.

DC and Paris, both among the world's most iconic capital cities, are prime case studies of how these worlds confront one another, yet still only a small part of the story. Studying music, or any locally oriented art form, helps us understand how vital the interpersonal connections are that undergird urban life as well as the music. As timeless as many of the recorded songs were, hardcore brought people together in a way that few other subcultures had. Gentrification has drawn lines, but punk's unique brand of disalienation became crucial to understanding and ultimately transcending those lines.

The goal of this book has been to reinforce the role that punk has continuously played in the development and escalation of popular counternarratives. Despite over 40 years of shifting ideologies and culturally embedded meanings, punk has remained as relevant and viable as it ever was. In its endurance through consistent reinvention, it provides a valuable chapter in the long and winding story of Franco-American cultural circulation. On a larger scale, it reinforces Yi Fu Tuan's (1977) thesis on how culture "strongly influences human behavior and values… but it overlooks the problem of shared traits that transcend cultural particularities and may therefore reflect the general human condition" (p. 5). However, this also points to a need for more sophisticated conceptions of music circulation, diffusion, and globalization among those who study urban life and people. Fortunately, however, punk is what you make of it, and it doesn't slow down or stop long enough for people to make complete sense of it.

The lopsided yet still symbiotic relationship between the musical undergrounds of DC and Paris illuminates dynamics of how the cities perceive one another, and how those perceptions change and ultimately, so do the cities. I chose to look at DC and Paris in particular because both are among the Western world's most symbolically powerful and powerfully symbolic cities, integral in the development of their respective countries' national identities while factoring into urban and social theory writ large. On the power of their respective national symbols and repositories of culture, both cities are among the world's biggest tourist magnets, and neither needs music tourism. Therefore, whatever tourism punk may encourage is that much more remarkable.

Regardless, both cities' punk histories, largely developed in comparably inauspicious circumstances and generally well outside the mainstream purview, have earned a level of legitimization that would have been unforeseeable for the scenes' progenitors who produced the songs, records, zines, and the shows. This retroactive public acceptance and curiosity has spread throughout the world, especially to other urban centers in Europe. In Berlin, everything from exhibits, coffee table books, and punk walking tours have been consistently available to visitors seeking to excavate music's role in the long-term punk hovel and flashpoint of Western and Eastern cultural life. Like their counterparts in DC, these presentations of culture allow the punk tourists to embrace icons from the past while engaging with the present and future of the music and its urban manifestations.

Though Mark Andersen (Positive Force DC) had been giving walking tours of the sites of DC punk history for years, other sources have been targeting tourists as well as the city's emergent young professional population. In 2015, the Travel & Leisure website, hardly a niche publication, posted an article promoting a DC Punk walking tour, hosted by Cynthia Connolly, through the DC Public Library's Punk Archive. So many people bought tickets; they could not even operate the tour logistically and had to move it inside. In DC and Paris, this post hoc sensationalism of "living" that era, absorbing that urban ethos in real time, may be impossible, but the emotional curiosity these cities have accrued over the past three decades in the wake of hardcore is remarkable. Guy Debord wrote in 1967 that we are living in the Society of the Spectacle; in many ways, he anticipated a wide swath of DC's identity (not to mention his own city's) 50 years on. The cumulative fascination with the hardcore era has also shed light on these geographic counternarratives and their meanings today. Thought the past has been, to quote Tuan (1977) rendered and made more accessible, questioning to whom these counternarratives are important and why they are important to those individuals is key. Circulation, like the music itself, has different meanings for different people, as do the cities.

This public ownership which music unwittingly encourages, though not always neatly consumed through mass-produced figurines or posters, can directly incorporate sense of place and elude the artist's original intention. Punk scenes, especially in DC, become inextricable from urban landscapes, and over time, these become commodified and sold. Though harDCore has not yet developed mass-marketing for tourists, locals in the Petworth neighborhood can order burgers named after Ian MacKaye, Dave Grohl, Henry Rollins, and Chuck Brown at the Satellite Room. On occasion, curious locals and visitors can bid on DC Punk walking tours led by Mark Andersen as occasional charity benefits.

The mainstream may not have been too interested for much of its early life cycle, but the scene was, collectively, fairly ingenious in keeping its stories and representations close to home. Sustainability went a long way in keeping these bands intimately tied to DC. It took decades for conglomerated popular culture to wrest some form of control from Minor Threat, and even then the most compromising by-product was unauthorized T-shirts at Urban Outfitters. As the years went by and the punks got older, they were able to maintain a relative amount of control over an uncontrollable narrative. I thought about this a lot in 2017 when Guy Picciotto, in town from Brooklyn and sitting near the court at a Wizards basketball game, appeared on the Jumbo-tron with his name and "Fugazi and Rites of Spring" underneath it. The screen-grab became a meme after Q and Not U's Chris Richards, by then the lead music writer for the *Washington Post*, tweeted it out. Moments like those are few and far between, but always funny. It doesn't matter how many people at the Wizards game knew who those bands were (or that they were even bands); his face on the big screen sent a strange message that they were irresponsible fans of Washington if they didn't. At the very least, it made for a humorous juxtaposition with the widely circulated image from *Instrument* of Guy dunking himself through a basketball hoop during a Philadelphia show in 1988. The same could be said for Canty's 2018 appearance on the sketch comedy show *Portlandia* alongside Henry Rollins and Krist Novoselic. All three came to the show at the behest of Fred Armisen, a longtime fan who came up playing in punk bands but rose to fame as a sketch comedian. Considering how antithetical harDCore was to popular culture, it's continuously surprising how pop culture tries catching up to harDCore nearly four decades later.

The same could be said for Paris' punk, oi!, and hardcore scenes. Philippe Roizès and friends from over the years have spent much of this decade telling their stories for the France Culture website, including podcasts setting the record straight about early skinhead culture. The blighted dynamics of 1980s Paris were slightly different from DC at the height of its crack era, but the retroactive, voyeuristic fascination with the era is strikingly similar for both. Expanding archives of underground music have been catalysts for those reinterpretations. Surviving members of L'Infanterie Sauvage and R.A.S. have reunited to practice, hang out, and give interviews their teenaged selves would never have imagined. Roizès opposes these reunions on principle because "reunions are not punk-rock," but the people in those bands were too important for him to not participate.

On the other hand, if harDCore ever had a directing ethos, it was accessibility. Many of the first few generations of DC punk took the sub-genre's founding ideology of uniting the performer and audience to a spatial extreme. Few branches of art embrace the idea of democracy and accessibility more actively than punk music, and DC has an urban ethos more inextricably linked with those ideals than anywhere. The case study of that scene's historically framed interaction with Paris, this underground cultural circulation, demonstrates how the global community reinforces that. This becomes even more important as "globalization" becomes a buzzword in American politics and those in power, seemingly unaware that punk ever happened, drive wedges between "Westernness" and "Americanness." Paul Adams (2007) reflected in the wake of the 2004 presidential election that Americans tend to view the world in black and white, and "the French are famous for attending to the grays" (p. 4). The political life of the twenty-first century has been, and will continue to be, "captured by the half-tones and shadows."

Furthermore, any static American imaginary of Europe is exceedingly harmful; teaching about Paris' violent era that birthed hardcore punk, as unsettling as those stories are, can be crucial in alerting outsiders to the city's realities. Too many Americans, internalizing unproblematic images propagated through mass media, fetishize EU countries as pockets of delicately preserved monolithic cultures, which is categorically untrue. The 2016 Brexit vote, similar to the 2016 American presidential election, leaned heavily on these idealized, exclusionary ethnic landscapes with the false idealism of "protecting" white culture from outside threats. This is unfair and barely different from French writers who never visited the US immortalizing the idea of America and images of Americans that often supplement each other. Watching any European national team can bring this to mind; even the Swiss national team has a diverse group of players. Isolationism does not work, and these success stories of international migration and multiculturalism are getting more prevalent and more visible. It is almost as if these imaginaries are ingrained (the tourism industry, a heavily modernist byproduct of World War II, is complicit in this) and it is up to both underground cultures and their nemesis the media (see Thornton 1996) to break down and diversify the global imaginary.

One outgrowth of punk is to help to disintegrate this colonialist world-view, and as Kevin Dunn (2016) wrote, provide "an interesting vehicle to explore the processes involved in contemporary globalization, some of globalization's contradictions, and the ways in which globalization is resisted and/or restructured" (p. 98). It also reminds us of the value of parity in urban revanchism and maintenance of the right to the city, because cities cannot reproduce themselves when turned into museum pieces; they must be changing constantly to meet the needs of their residents. Their citizens and the cities themselves must be able to live in the modern world and adapt to its changes.

The landscape counternarratives associated with underground music call attention to the need for livable cities for both those engaging with the neoliberal, capitalist engine as well as those who want to operate (even in their leisure time) on its fringes or way outside. In the epoch of its emergence, punk may have problematized the dichotomy between urban art and commerce. Understanding the circulatory nature of punk as well as underground music in general has further nuanced our understanding of numerous dichotomies: high culture versus low culture, heritage tourism versus package tourism, and folk culture versus the popular. It continues to do so, and these urban backdrops and the people who propel the circulations of the city remain at the heart of the process.

Bibliography

Adams, P. C. (2007). *Atlantic reverberations: French representations of an American presidential election.* London: Ashgate.

Dunn, K. (2016). *Global punk: Resistance and rebellion in everyday life.* New York: Bloomsbury.

Thornton, S. (1996). *Club cultures: Music, media, and subcultural capital.* Middletown: Wesleyan University Press.

Tuan, Y. (1977). *Space and place: The perspective of experience* (10th ed.). Rochester: University of Minnesota Press.

Bibliography

Adams, P. C. (2007). *Atlantic reverberations: French representations of an American presidential election*. London: Ashgate.

Alderman, D. H. (2002). Writing on the Graceland wall: On the importance of authorship in pilgrimage landscapes. *Tourism Recreation Research, 27*(2), 27–33.

Allison, F. H. (2010). Remembering a Vietnam war firefight: Changing perspectives over time. In R. Perks & A. Thompson (Eds.), *The oral history reader* (2nd ed., pp. 68–78). London: Routledge.

Andersen, M., & Jenkins, M. (2001). *Dance of days: Two decades of punk in the Nation's capital*. New York: Akashic Books.

Angrosino, M. (2008). *Exploring oral history: A window on the past*. Long Grove: Waveland Press.

Augenstein, N. (2016, October 18). *'Startling': Ian MacKaye reacts to Bad Brains' Rock Hall of Fame nomination*. Retrieved February 19, 2017, from http://wtop.com/music/2016/10/startling-ian-mackaye-reacts-to-bad-brains-rock-hall-of-fame-nomination/

Azerrad, M. (2001). *Our band could be your life: Scenes from the American indie underground, 1981–1991*. New York: Little, Brown.

Bad Brains. (1982). Banned in DC. On *Bad Brains* [Compact Disc]. New York: ROIR.

Baudrillard, J. (2006). *Utopia deferred: Writings from Utopie (1967–1978)* (S. Kendall, Trans.). New York: Semiotext.

Bauman, Z. (1994). Desert spectacular. In K. Tester (Ed.), *The Flâneur* (pp. 138–161). London: Routledge.

Baxter, J., & Eyles, J. (1997). Evaluating qualitative research in social geography: Establishing 'rigour' in interview analysis. *Transactions of the Institute of British Geographers, 22*(4), 505–525.

Beauchez, J. (2014). La rue comme héroïne: expériences punk et skinhead en France. *Anthropologica, 56*(1), 193–204.

Becker, H. S. (1967). History, culture and subjective experience: An exploration of the social bases of drug-induced experiences. *Journal of Health and Social Behavior, 8*(3), 163–176.

Beeber, S. L. (2014). Juidos 'n' Decaf Italians: Irony, Blasphemy, and Jewish Shtick. In B. Lashua, K. Spracklen, & S. Wagg (Eds.), *Sounds and the city: Popular music, place and globalization* (pp. 76–91). London: Palgrave Macmillan.

Bell, T. (1998). Why Seattle? An examination of an alternative rock culture hearth. *Journal of Cultural Geography, 18*(1), 35–47.

Benjamin, W. (1999). *The Arcades project* (H. Eiland & K. McLaughlin, Trans.). Cambridge, MA: Harvard University Press.

Berland, J. (1992). Angels dancing: Cultural technologies and the production of space. In L. Grossberg & N. Cary (Eds.), *Cultural studies* (pp. 38–51). New York: Routledge.

Blacking, J. (1974). *How musical is man?* Seattle: University of Washington Press.

Blee, K. (2010). Evidence, empathy, and ethics: Lessons from oral histories of the Klan. In R. Perks & A. Thompson (Eds.), *The oral history reader* (2nd ed., pp. 322–331). London: Routledge.

Bottà, G. (2009). The city that was creative and did not know. *European Journal of Cultural Studies, 12*(3), 349–365.

Bourdieu, P. (1984). *Distinction: A social critique of the judgment of taste.* Trans. Nice, Richard. Cambridge, MA: Harvard University Press.

Bourgeois, A. (2016). Email correspondence, 21 Sept.

Boutros, A., & Straw, W. (2010). Introduction. In A. Boutros & W. Straw (Eds.), *Circulation and the city: Essays on urban culture* (pp. 3–21). Montreal: McGill-Queen's University Press.

Brannon, N. (2007). *The anti-matter anthology: A 1990s post-punk & hardcore reader.* Huntington Beach: Revelation Records.

Brennan-Horley, C., Connell, J., & Gibson, C. (2007). The Parkes Elvis revival festival: Economic development and contested place identities in rural Australia. *Geographical Research, 45*(1), 71–84.

Briggs, J. (2015). *Sounds French: Globalization, cultural communities, and pop music, 1958–1980*. Oxford: Oxford University Press.

Bright, C. F., & Butler, D. L. (2015). Webwashing the tourism plantation. In S. P. Hanna, A. E. Potter, E. A. Modlin, P. Carter, & D. L. Butler (Eds.), *Social memory and heritage tourism methodologies* (pp. 31–43). London: Routledge.

Buck-Morss, S. (1986). The flaneur, the sandwichman and the whore: The politics of loitering. *New German Critique, 39*, 99–140.

Burman, J. (2010). Absence, "removal," and the everyday life in the Diasporic City: Anti-detention/deportation activism in Montreal. In A. Boutros & W. Straw (Eds.), *Circulation and the city: Essays on urban culture* (pp. 99–117). Montreal: McGill-Queen's University Press.

Burton, S. K. (2010). Issues in cross-cultural interviewing: Japanese women in England. In R. Perks & A. Thompson (Eds.), *The oral history reader* (2nd ed., pp. 166–183). London: Routledge.

Cadiot, P. (2015). Email correspondence, 21 Aug.

Cameron, J. (2010). Focusing on the focus group. In I. P. Hay (Ed.), *Qualitative research methods in human geography* (pp. 152–172). Don Mills: Oxford University Press Canada.

Carter, P. (2015). Virtual ethnography: Placing emotional geographies via YouTube. In S. P. Hanna, A. E. Potter, E. A. Modlin, P. Carter, & D. L. Butler (Eds.), *Social memory and heritage tourism methodologies* (pp. 48–67). London: Routledge.

Casana, M. (2015). Email correspondence, 23 Aug.

Cathcart, R. S. (1972). New approaches to the study of movements: Defining movements rhetorically. *Western Speech, 36*(2), 82–88.

Chambers, I. (1997). Maps, movies, musics and memory. In D. B. Clarke (Ed.), *The cinematic city* (pp. 230–240). London: Routledge.

Charbonnier, M. (2016). Email correspondence, 1 Feb.

Chhabra, D., Healy, R., & Sills, E. (2003). Staged authenticity and heritage tourism. *Annals of Tourism Research, 30*(3), 702–719.

Clerval, A. (2008). *La Gentrification À Paris Intra-Muros: Dynamiques Spatiales, Rapports Sociaux Et Politiques Publiques*. Paris: Université Panthéon-Sorbonne-Paris I.

Cockburn, P. (1996, October 22). 'This is a provocation. Do you want me to get on my plane and go back to France?' Furious Chirac threatens to cut short visit to Israel after clash with security forces. *The Independent*, p. 15. Retrieved from http://www.independent.co.uk/news/world/this-is-a-provocation-do-you-want-me-to-get-on-my-plane-and-go-back-to-france-1359705.html

Cohen, S. (1995). Sounding out the city: Music and the sensuous production of place. *Transactions of the Institute of British Geographers, 20*(4), 434–446.

Cohen, S. (2005). Country at the heart of the city: Music, heritage, and regeneration in Liverpool. *Ethnomusicology, 49*(1), 25–48.

Cohen, S., Schofield, J., & Lashua, B. (2010). Introduction to the special issue: Music, characterization and urban space. *Popular Music History, 4*(2), 105–110.

Connell, J., & Gibson, C. (2004). Vicarious journeys: Travels in music. *Tourism Geographies, 6*(1), 2–25.

Connell, J., & Gibson, C. (2008). 'No passport necessary': Music, record covers and vicarious tourism in post-war Hawai'i. *Journal of Pacific History, 43*(1), 51–75.

Connolly, C. (2016). Personal conversation, 22 July. Arlington, VA.

Connolly, C., Clague, L., & Cheslow, S. (1988). *Banned in DC: Photos and anecdotes from the DC punk underground* (7th ed., pp. 79–85). Washington, DC: Sundog Publications.

Corkery, C. K., & Bailey, A. J. (1994). Lobster is big in Boston: Postcards, place commodification, and tourism. *GeoJournal, 34*(4), 491–498.

Cosgrove, D. (1989). Geography is everywhere: Culture and symbolism in human landscapes. In D. Gregory & R. Walford (Eds.), *Horizons in human geography* (pp. 118–135). New York: Barnes & Noble Books.

Crang, M. (1996). Envisioning urban histories: Bristol as palimpsest, postcards, and snapshots. *Environment and Planning A, 28*(3), 429–452.

Crawford, S. (Writer). (2015). *Salad days: A decade of Punk in Washington, DC (1980–1990)*. New Rose Films.

Crossley, N. (2015). Totally wired: The network of structure of the post-punk world of Liverpool, Manchester and Sheffield 1976–80. In N. Crossley, S. McAndrew, & P. Widdop (Eds.), *Social networks and music worlds* (pp. 40–60). London: Routledge.

Crossley, N., McAndrew, S., & Widdop, P. (2015). Introduction. In N. Crossley, S. McAndrew, & P. Widdop (Eds.), *Social networks and music worlds* (pp. 1–13). London: Routledge.

Culli, D. R. (2004). *"Never could read no road map": Geographic perspectives on the grateful dead*. PhD Book, Southern Illinois University, Edwardsville.

Cuzin, É. (2016, August 13). L'âme punk de Washington résonne encore. *La Presse*. Retrieved from http://www.lapresse.ca/arts/musique/201604/28/01-4976060-lame-punk-de-washington-resonne-encore.php

D'Angelo, P. J. (2000). Allmusic.com. Review of *Flash Flash Flash* by The Explosion. Retrieved April 17, 2017, from http://www.allmusic.com/album/flash-flash-flash-mw0000608751

Dann, G. M. (1977). Anomie, ego-enhancement and tourism. *Annals of Tourism Research, 4*(4), 184–194.

Dauvillier, G. (2015). Personal interview, 6 July. Montreuil, France.

Davies, H. (2001). All rock and roll is homosocial: The representation of women in the British rock music press. *Popular Music, 20*(03), 301–319.

de Tocqueville, A. (1978). *Democracy in America* (1835, 21st ed.). New York: Mentor.

Delevacque, S. (2015). Personal interview, 3 July. Paris, France.

DeLyser, D. (2001). "Do you really live here?" thoughts on insider research. *Geographical Review, 91*(1/2), 441–453.

Domosh, M. (1998). Geography and gender: Home, again? *Progress in Human Geography, 22*(2), 276–282.

Domosh, M., & Seager, J. (2001). *Putting women in place: Feminist geographers make sense of the world*. New York: The Guilford Press.

Dos Santos, D. G. (2015). Between drums and drones: The urban experience in Sao Paulo's punk music. In P. Guerra & T. Moreira (Eds.), *Keep it simple, make it fast: An approach to underground music scenes* (Vol. 1, pp. 135–145). Porto: Universidade do Porto – Faculdade de Letras.

Duncombe, S. (2008). *Notes from underground: Zines and the politics of alternative culture*. New York: Microcosm Publishing.

Dunn, K. (2010). Interviewing. In I. P. Hay (Ed.), *Qualitative research methods in human geography* (pp. 101–137). Don Mills: Oxford University Press Canada.

Dunn, K. (2013). One punk's guide to Indonesia. *Razorcake, 76*, 34–45.

Dunn, K. (2016a). Email correspondence, 16 June.

Dunn, K. (2016b). *Global punk: Resistance and rebellion in everyday life*. New York: Bloomsbury.

Dwyer, O. J. (2004). Symbolic accretion and commemoration. *Social & Cultural Geography, 5*(3), 419–435.

Earles, A. (2014). *Gimme Indie Rock: 500 essential American underground rock albums 1981–1996*. Minneapolis: Voyageur Press.

Easley, D. B. (2015). Riff schemes, form, and the genre of early American hardcore punk (1978–83). *Music Theory Online, 21*(1), 1–21.

Edensor, T. (2009). Tourism. In R. Kitchin & N. Thrift (Eds.), *International encyclopedia of human geography* (Vol. 11, pp. 301–312). Oxford: Elsevier.

Estanove, L. (2015, March 28). *Postcard records: Glasgow's post-punk myth?* Paper presented at the disorder: Histoire Sociale des mouvements punk/post-punk, Paris.

Fairchild, C. (1995). "Alternative"; music and the politics of cultural autonomy: The case of Fugazi and the D.C. Scene. *Popular Music and Society, 19*(1), 17–35.

Feinberg, P. (2007). Bad religion punk rock Greg Graffin on his scholarly gig at UCLA. *UCLA Magazine.* Retrieved April 17, 2017, from http://magazine. ucla.edu/exclusives/bad-religion_greg-graffin/

Feld, S. (2012). *Sound and sentiment: Birds, weeping, poetics and song in Kaluli expression.* Durham: Duke University Press.

Firminhac, O., Samant, C., & Lobert, C. (2015). Group interview, 24 July. Marseille, France.

Flanagan, W. G. (1993). *Contemporary urban sociology.* Cambridge: Cambridge University Press.

Florida, R. (2002). *The rise of the creative class.* New York: Basic Books.

Foley, M. S. (2015). *Dead Kennedy's fresh fruit for rotting vegetables. 33 1/3* (Vol. 105). New York: Bloomsbury.

Foreman, G. (2015). Guy Picciotto. *Psychic Gloss.* Retrieved November 11, 2016, from http://www.psychicgloss.com/articles/6028

Foucault, M. (1972). *The archaeology of knowledge: Translated from the French by AM Sheridan Smith.* New York: Pantheon Books.

Fournier, K. (2016). Nazi signifiers and the narrative of class warfare in British punk. In M. M. Hall, S. Howes, & C. M. Shahan (Eds.), *Beyond no future: Cultures of German punk* (pp. 91–108). New York: Bloomsbury.

Freelunsch, A. (1986). Europe: Ahoy Matey! *Maximumrocknroll*, p. 34.

Fremaux, S., & Fremaux, M. (2013). Remembering the Beatles' legacy in Hamburg's problematic tourism strategy. *Journal of Heritage Tourism, 8*(4), 303–319.

Freud, C. (1959). Portrait of the beatnik. *Encounters, 12*(6), 42–46.

Frith, S. (1996). Music and identity. In S. Hall & P. D. Gay (Eds.), *Questions of cultural identity* (pp. 108–127). London: Sage.

Fuchs, C. (1999). Images. In B. Horner & T. Swiss (Eds.), *Key terms in popular music and culture* (pp. 178–187). Malden: Blackwell Publishing.

Fugazi. (1989). Suggestion. On *13 Songs* [Compact Disc]. Washington, DC: Dischord Records.

Gallagher, M., & Prior, J. (2014). Sonic geographies: Exploring phonographic methods. *Progress in Human Geography, 38*(2), 267–284.

Gallois, L. (1923). The origin and growth of Paris. *Geographical Review, 13*(3), 345–367.

Gandy, M. (2016, April 13). *Cultural geographies annual lecture: Urban Atmospheres.* Paper presented at the Association of American Geographers Annual Meeting, San Francisco.

Garrett, B. L. (2010). Urban explorers: Quests for myth, mystery and meaning. *Geography Compass, 4*(10), 1448–1461.

Garrett, B. L. (2011). Videographic geographies: Using digital video for geographic research. *Progress in Human Geography, 35*(4), 521–541.

Gay, L. C. (2003). Before the deluge: The technoculture of song-sheet publishing viewed from late-nineteenth-century Galveston. In R. T. Lysloff & L. C. Gay (Eds.), *Music and technoculture* (pp. 204–232). Middletown: Wesleyan University Press.

Geertz, C. (1973). *The interpretation of cultures: Selected essays* (Vol. 5019). New York: Basic Books.

Gelézeau, M. (2015). Personal interview, 12 July. Paris, France.

Gibbs, T. (2012). Brendan Canty interview. *One Week/One Band Tumblog.* Retrieved from http://oneweekoneband.tumblr.com/post/11690077029/brendan-canty-interview

Gibson, C., & Connell, J. (2007). Music, tourism and the transformation of Memphis. *Tourism Geographies, 9*(2), 160–190.

Gibson, C., & Davidson, D. (2004). Tamworth, Australia's 'country music capital': Place marketing, rurality, and resident reactions. *Journal of Rural Studies, 20*(4), 387–404.

Gillen, J. (2012). Investing in the field: Positionalities in money and gift exchange in Vietnam. *Geoforum, 43*(6), 1163–1170.

Gillett, C. (1970). *The sound of the city: The rise of rock and roll.* New York: Outerbridge & Dienstfrey.

Gillette, H. (2016). Introduction: For a city in transition, questions of social justice and economic viability remain. In D. Hyra & S. Prince (Eds.), *Capital Dilemma: Growth and inequality in Washington, DC* (pp. 1–8). London: Routledge.

Goode, J. M. (2003). *Capital losses: A cultural history of Washington's destroyed buildings.* Washington, DC: Smithsonian Books.

Gotham, K. F. (2005). Tourism gentrification: The case of New Orleans' vieux carre (French quarter). *Urban Studies, 42*(7), 1099–1121.

Gramsci, A. (2012). *Selections from the Prison Notebooks* (Q. Hoare & G. N. Smith, Trans.). New York: International Publishers.

Green, N. L. (2014). *The other Americans in Paris: Businessmen, countesses, wayward youth, 1880–1941*. Chicago: University of Chicago Press.

Greenberg, R. (2016, July 27). Ryan Adams covered Fugazi and confronted a heckler at the Lincoln Theatre in D.C. last night. *Washington Post Express*. Retrieved from https://www.washingtonpost.com/express/wp/2016/07/27/ryan-adams-covered-fugazi-and-confronted-a-heckler-at-the-lincoln-theatre-in-d-c-last-night/?utm_term=.e52341497b4c

Gresser, N. (2015). Personal interview, 11 July. Paris, France.

Gumprecht, B. (1998). Lubbock on everything: The evocation of place in popular music (a West Texas example). *Journal of Cultural Geography, 18*(1), 61–81.

Gunderman, H. (2013). *"The music never stopped": Naming business as a method for memorializing the legacy of the grateful dead*. Laramie: Master of Arts, University of Wyoming.

Gunderman, H. C., & Harty, J. P. (2016). "The music never stopped": Naming businesses as a method for remembering the grateful dead. *Journal of Cultural Geography, 3*, 1–23.

Guralnick, P. (1998). *Searching for Robert Johnson*. New York: Plume Books.

Haenfler, R. (2004). Collective identity in the straight edge movement: How diffuse movements foster commitment, encourage individualized participation, and promote cultural change. *The Sociological Quarterly, 45*(4), 785–805.

Hagerman, C. (2007). Shaping neighborhoods and nature: Urban political ecologies of urban waterfront transformations in Portland, Oregon. *Cities, 24*(4), 285–297.

Hall, S. (1968). *The hippies: An American 'moment'*. Birmingham: Centre for Contemporary Cultural Studies, University of Birmingham.

Hall, M. M. (2016). Cold wave: French post-punk fantasies of Berlin. In M. M. Hall, S. Howes, & C. M. Shahan (Eds.), *Beyond no future: Cultures of German punk* (pp. 149–166). New York: Bloomsbury.

Hall, M. M., Howes, S., & Shahan, C. M. (2016a). German Mania: A Coda. In M. M. Hall, S. Howes, & C. M. Shahan (Eds.), *Beyond no future: Cultures of German punk* (pp. 167–170). New York: Bloomsbury.

Hall, M. M., Howes, S., & Shahan, C. M. (2016b). Punk matters: An introduction. In M. M. Hall, S. Howes, & C. M. Shahan (Eds.), *Beyond no future: Cultures of German punk* (pp. 1–15). New York: Bloomsbury.

Hebdige, D. (1979). *Subculture: The meaning of style*. New York: Routledge.

Hebdige, D. (2012). Contemporizing 'subculture': 30 years to life. *European Journal of Cultural Studies, 15*(3), 399–424.

Henderson, F. M. (1974). The image of New York City in American popular music: 1890–1970. *New York Folklore Quarterly, 30*, 267–279.

Herbert, S. (2000). For ethnography. *Progress in Human Geography, 24*(4), 550–568.

Herbert, S. (2010). A taut rubber band: Theory and empirics in qualitative geographic research. In D. DeLyser, S. Herbert, S. Aitken, M. Crang, & L. McDowell (Eds.), *The Sage handbook of qualitative geography* (pp. 69–81). London: Sage.

Hernandez-Sang, V. (2016, March 4). *"All are welcome in our band": Pan-Latino and inclusive social cohorts in Washington DC's Latin (American Popular Dance) music scene*. Paper presented at the Society for Ethnomusicology, Southeast and Caribbean Chapter Annual Meeting, San Fernando, Trinidad & Tobago.

Herzfield, M. (1996). Productive discomfort: Anthropological fieldwork and the dislocation of etiquette. In *Field work: Sites in literacy and cultural studies* (pp. 41–51). New York: Routledge.

Herzock, N. (2015). Personal interview, 15 July. Malakoff, France.

Hills, M. (2012). "Proper distance" in the ethical positioning of scholar-fandoms: Between academics. In K. Larson & L. Zubernis (Eds.), *Fan culture: Theory/practice* (pp. 14–37). Newcastle: Cambridge Scholars.

Hoelscher, S. (2009). Landscape iconography. In N. Thrift & R. Kitchin (Eds.), *International encyclopedia of human geography* (pp. 132–139). Amsterdam: Elsevier.

Hood, M. (1960). The challenge of "bi-musicality". *Ethnomusicology, 4*(2), 55–59.

Hopkinson, N. (2012). *Go-go live: The musical life and death of a Chocolate City*. Durham: Duke University Press.

Hough, E. (2011). Rethinking authenticity and tourist identity: Expressions of territoriality and belonging among repeat tourists on the Greek island of Symi. *Journal of Tourism and Cultural Change, 9*(2), 87–102.

Hracs, B. J., Jakob, D., & Hauge, A. (2013). Standing out in the crowd: The rise of exclusivity-based strategies to compete in the contemporary marketplace for music and fashion. *Environment and Planning A, 45*(5), 1144–1161.

Hracs, B., Virani, T. E., & Seman, M. (Eds.). (2016). *The production and consumption of music in the digital age*. London: Routledge.

Hudson, R. (2006). Regions and place: Music, identity and place. *Progress in Human Geography, 30*(5), 626–640.

Hyra, D., & Prince, S. (2016). Preface. In D. Hyra & S. Prince (Eds.), *Capital dilemma: Growth and inequality in Washington, DC* (pp. xiii–xxvi). London: Routledge.

Inglis, I. (2012). *The Beatles in Hamburg*. Reverb. London: Reaktion Books.

Jacobs, J. (1958). Downtown is for people. In Editors of Fortune (Ed.), *The exploding metropolis* (pp. 140–168). Garden City: Doubleday.

Jacobsen, H. N. (Ed.). (1965). *A guide to the architecture of Washington, D.C.* New York: Frederick A. Praeger.

Jaskowski, R. (2015). Personal interview, 28 July. Malakoff, France.

Johansson, O., & Bell, T. (Eds.). (2009). *Sound, society, and the geographies of popular music*. Farnham: Ashgate.

John, D. (2016). *July 12, 1979: The night disco died or didn't*. Retrieved July 16, 2016, from http://www.npr.org/2016/07/16/485873750/july-12-1979-the-night-disco-died-or-didnt

Johnson, M. (2006, September 15). A chateau fit for a president. *The New York Times Online*. Retrieved from http://www.nytimes.com/2006/09/15/opinion/15iht-edjohnson.html

Kane, M. K. (2016). Email correspondence, 22 July.

Kassabian, A. (1999). Popular. In B. Horner & T. Swiss (Eds.), *Key terms in popular music and culture* (pp. 113–123). Malden: Blackwell Publishing.

Keeffe, G. (2010). Compost city: Underground music, collapsoscapes and urban regeneration. *Popular Music History, 4*(2), 145–159.

Kelly, J. (2014). *Fugazi, a standout from D.C.'s musical past, pops up in an unexpected place*. Retrieved from https://www.washingtonpost.com/local/fugazi-a-standout-from-dcs-musical-past-pops-up-in-an-unexpected-place/2014/12/28/055f260c-8eae-11e4-a900-9960214d4cd7_story.html?utm_term=.5f9aa2d0d4cc

Kelly, J. (2016, August 1). If these walls could talk, they'd probably scream. *The Washington Post Online*. Retrieved from https://www.washingtonpost.com/local/if-these-walls-could-talk-theyd-probably-scream/2016/08/01/86bbed62-5751-11e6-831d-0324760ca856_story.html?utm_term=.26f1de1fea31

Kelly, M., Jones, T., & Forbes, J. (1995). Modernization and Avant-gardes. In J. Forbes & M. Kelly (Eds.), *French cultural studies* (pp. 140–182). Oxford: Oxford University Press.

King, S. A. (2006). Memory, mythmaking, and museums: Constructive authenticity and the primitive blues subject. *Southern Communication Journal, 71*(3), 235–250.

Kitchin, R., & Tate, N. (2013). *Conducting research in human geography: Theory, methodology and practice.* London: Routledge.

Klemek, C. (2016). Exceptionalism and the National Capital in late 20th-century Paris and Washington, DC. In D. Hyra & S. Prince (Eds.), *Capital dilemma: Growth and inequality in Washington, DC* (pp. 11–26). London: Routledge.

Knupp, R. E. (1981). A time for every purpose under heaven: Rhetorical dimensions of protest music. *Southern Speech Communication Journal, 46*(4), 377–389. https://doi.org/10.1080/10417948109372503.

Konan, A. (2016, August 10). Black Dragons: The black punk gang who fought racism & skinheads in 1980s France. *Okay Africa.* Retrieved from http://www.okayafrica.com/featured/black-punk-black-dragons-france/

Kong, L. (1995). Popular music in geographical analyses. *Progress in Human Geography, 19*(2), 183–198.

Krims, A. (2007). *Music and urban geography.* London: Routledge.

Krüger, S. (2014). Branding the city: Music tourism and the European Capital of culture event. In S. Kruger & R. Trandafoiu (Eds.), *The globalization of musics in transit: Music migration and tourism* (pp. 135–159). London: Routledge.

Krüger, S., & Trandafoiu, R. (2014). Introduction: Touristic and migrating musics in transit. In S. Krüger & R. Trandafoiu (Eds.), *The globalization of musics in transit: Music migration and tourism* (pp. 1–33). London: Routledge.

Krulik, J. (2013). Email correspondence, 25 Oct.

Kruse, H. (1993). Subcultural identity in alternative music culture *Popular Music, 12*(1), 33–41.

Kruse, R. J. (2005). The Beatles as place makers: Narrated landscapes in Liverpool, England. *Journal of Cultural Geography, 22*(2), 87–114.

Kugelberg, J., & Vermès, P. (Eds.). (2011). *Beauty is in the street: A visual record of the May 68 Paris uprising.* London: Four Corners Books.

Le Roux, F. (2015). Personal interview, 23 July. Rouen, France.

Leaver, D., & Schmidt, R. A. (2009). Before they were famous: Music-based tourism and a musician's hometown roots. *Journal of Place Management and Development, 2*(3), 220–229.

Lee, S. (2005). Punk "noir": Anarchy in two idioms. *Yale French Studies,* 177–188.

Ley, D., & Cybriwsky, R. (1974). Urban graffiti as territorial markers. *Annals of the Association of American Geographers, 64*(4), 491–505.

Long, P. (2014). Popular music, psychogeography, place identity and tourism: The case of Sheffield. *Tourist Studies, 14*(1), 48–65.

Longhurst, R. (2009). YouTube: A new space for birth? *Feminist Review, 93,* 46–63.

Longhurst, R. (2010). Semi-structured interviews and focus groups. In N. Clifford, S. French, & G. Valentine (Eds.), *Key methods in geography* (2nd ed., pp. 103–115). London: Sage.

Looseley, D. (2005). Fabricating Johnny: French popular music and national culture. *French Cultural Studies, 16*(2), 191–203.

Loughran, M. E. (2008). *Community powered resistance: Radio, music scenes and musical activism in Washington, D.C.* PhD, Brown University, Providence.

Lukinbeal, C. (1998). Reel-to-real urban geographies: The top five cinematic cities in North America. *The California Geographer, 38*(1), 64–78.

Lukinbeal, C. (2014). Geographic media literacy. *Journal of Geography, 113*(2), 41–46.

Lummis, T. (2010). Structure and validity in oral evidence. In R. Perks & A. Thompson (Eds.), *The oral history reader* (2nd ed., pp. 255–260). London: Routledge.

MacKaye, I. (2016). Skype interview, 13 Aug.

MacKaye, I., Picciotto, G., Canty, B., & Lally, J. (2001). Cashout. Prod. Zientara, Don and Fugazi. *The Argument.* Washington, DC: Dischord Records.

Maimone, H. (2016). Email correspondence, 7 July.

Mallinder, S. (2013). Sounds incorporated: Dissonant sorties into popular music. In M. Goddard, B. Halligan, & N. Spelman (Eds.), *Resonances: Noise and contemporary music* (pp. 81–94). Los Angeles: Bloomsbury.

Marcus, S. (2010). *Girls to the front: The true story of the riot grrrl revolution.* New York: Harper Collins.

Markley, S., & Sharma, M. (2016). Keeping Knoxville scruffy?: Urban entrepreneurialism, creativity, and gentrification down the urban hierarchy. *Southeastern Geographer, 56*(4), 384–408.

Maskell, S. (2009). Performing punk: Bad brains and the construction of identity. *Journal of Popular Music Studies, 21*(4), 411–426.

Masserman, J. H. (1967). The beatnik: Up—, down—, and off—. *Archives of General Psychiatry, 16*(3), 262–267.

Massey, D. (1994). *Place, space and gender.* Minneapolis: University of Minnesota Press.

McCloud, S. (2016). Email correspondence, 4 Aug.

McLuhan, M. (1964). *Understanding media: The extensions of man.* New York: Signet.

Médioni, G. (2007). *30 Ans de Rock Français: de Telephone a Dionysos.* Paris: L'Archipel.

Mension, N. (2015). Email correspondence, 13 Sept.

Merriam, A. P. (1964). *The anthropology of music.* Boston: Northwestern University Press.

Millan, G., Rigby, B., & Forbes, J. (1995). Industrialization and its discontents (1870–1944). In J. Forbes & M. Kelly (Eds.), *French cultural studies* (pp. 11–53). Oxford: Oxford University Press.

Miltzine, G. (1983). The history of French punk. *Flipside*, p. 38.

Minuchin, L. (2014). Noise, language, and public protest: The cacerolazos in Buenos Aires. In M. Gandy & B. Nilsen (Eds.), *The acoustic city* (pp. 201–205). London: Jovis.

Mitchell, T. (1996). *Popular music and local identity: Rock, pop, and rap in Europe and Oceania.* Leicester: Leicester University Press.

Modell, J. (2011). *Emergency & I* (The Dismemberment Plan) Reissue LP Liner Notes. Portland: Barsuk Records.

Molotch, H. (1976). The city as a growth machine: Toward a political economy of place. *American Journal of Sociology, 82*, 309–332.

Moreau, T., & Alderman, D. H. (2011). Graffiti hurts and the eradication of alternative landscape expression. *Geographical Review, 101*(1), 106–124.

Morley, P. (2005). *Words and music: A history of pop in the shape of a city.* Athens: University of Georgia Press.

Morrison, T. (1999). Spider in the Snow (The Dismemberment Plan). On *Emergency & I* [Compact Disc]. Washington, DC: DeSoto Records.

Moulard, J.-F. (2016). Email correspondence, 19 Feb.

Negus, K. (1999). *Music genres and corporate cultures.* London: Routledge.

Nelson, V. (2013). *An introduction to the geography of tourism.* Lanham: Rowman & Littlefield.

Nettl, B. (2005). *The study of ethnomusicology: Thirty-one issues and concepts.* Urbana: University of Illinois Press.

Nettleford, R. M. (1970). *Mirror, mirror: Identity, race, and protest in Jamaica.* Kingston: W. Collins and Sangster.

Nevarez, L. (2013). How joy division came to sound like Manchester: Myth and ways of listening in the neoliberal city. *Journal of Popular Music Studies, 25*(1), 56–76.

Niceley, T. (2016). Email correspondence, 10 Aug.

No Author Credited. (1965). *A guide to the architecture of Washington, D.C.* New York: Frederick A. Praeger.

No Author Credited. (1977, April 9). Jazz Reissues become growing industry. *Billboard Magazine*, p. F-5.

Novak, D. (2013). *Japanoise: Music at the edge of circulation.* Durham: Duke University Press.

O'Connell, J. (2015, August 18). D.C. sets tourism record with 20 million visitors. *The Washington Post.* Retrieved from https://www.washingtonpost.com/news/digger/wp/2015/08/18/d-c-sets-tourism-record-with-20-million-visitors/

O'Connor, A. (2008). *Punk record labels and the struggle for autonomy: The emergency of DIY.* Lanham: Lexington Books.

Ohashi, S. (2012). Classification of paradigms and approaches in the present tourism research. In *Academic world of tourism studies* (Vol. 1, pp. 9–17). Wakayama: Wakayama University.

Olsen, M. (1998). Everybody loves our town: Scenes, spatiality, migrancy. In T. Swiss, J. Sloop, & A. Herman (Eds.), *Mapping the beat: Popular music and contemporary theory.* London: Wiley-Blackwell.

Pappalardo, A. (2014). Why Fugazi are still the best punk band in the world – An Op-Ed. *Alternative Press.* Retrieved from http://www.altpress.com/features/entry/fugazi_are_the_best_punk_band_in_the_world

Perrauld, M. (2015). Personal interview, 25 July. Lyon, France.

Picciotto, G. (2016). Email correspondence, 16 Aug.

Pinto, A. C. (2015). Aesthetics, anti-aesthetics and "bad taste": A brief journey through Portuguese punk record covers (1977–1998). In P. Guerra & T. Moreira (Eds.), *Keep it simple, make it fast: An approach to underground music scenes* (Vol. 1, pp. 107–123). Porto: Universidade do Porto – Faculdade de Letras.

Pons, F., Ortega, N., & Casseville, G. (2015). Group interview, 26 July. Lyon, France.

Porcello, T. (2005). Music mediated as live in Austin: Sound technology and recording practice. In P. D. Greene & T. Porcello (Eds.), *Wired for sound: Engineering and technologies in sonic cultures* (pp. 103–117). Middletown: Wesleyan University Press.

Pothier, B. (2015). Personal interview, 8 July. Paris, France.

Prattichizzo, G. (2015). Social media is the new punk. User experience, social music and DIY culture. In P. Guerra & T. Moreira (Eds.), *Keep it simple, make it fast: An approach to underground music scenes* (Vol. 1, pp. 313–327). Porto: Universidade do Porto – Faculdade de Letras.

Préteceille, E. (2007). Is gentrification a useful paradigm to analyse social changes in the Paris metropolis? *Environment and Planning A, 39*(1), 10–31.

Raboud, P. (2015). DIY culture and youth struggles for autonomy in Switzerland: Distortion of the punk scene. In P. Guerra & T. Moreira (Eds.), *Keep it simple, make it fast: An approach to underground music scenes* (Vol. 1, pp. 29–36). Porto: Universidade do Porto – Faculdade de Letras.

Regis, H. A., & Walton, S. (2008). Producing the folk at the New Orleans Jazz and heritage festival. *Journal of American Folklore, 121*(482), 400–440.

Reia, J. (2015). I've got straight edge: Discussions on aging and gender in an underground musical scene. In P. Guerra & T. Moreira (Eds.), *Keep it simple, make it fast: An approach to underground music scenes* (Vol. 1, pp. 125–134). Porto: Universidade do Porto – Faculdade de Letras.

Rimmer, N. V. (2016). Email correspondence, 8 June.

Roizès, P. (2015). Personal interview, 3 July. Paris, France.

Roizès, P. (2016). Email correspondence, 23 June.

Rose, G. (1993). *Feminism & geography: The limits of geographical knowledge.* Hibbing: University of Minnesota Press.

Rose, G. (1997). Situating knowledges: Positionality, reflexivities and other tactics. *Progress in Human Geography, 21*(3), 305–320.

Rose, G. (2012). *Visual methodologies: An introduction to researching with visual materials* (3rd ed.). London: Sage.

Sauer, C. O. (1925). The morphology of landscape. *University of California Publications in Geography, 2*(2), 19–53.

Schroeder, C. G. (2014). (Un)holy Toledo: Intersectionality, interdependence, and neighborhood (trans)formation in Toledo, Ohio. *Annals of the Association of American Geographers, 104*(1), 166–181.

Scriven, M., Hewitt, N., Kelly, M., & Atack, M. (1995). War and class wars (1914–1944) In J. Forbes & M. Kelly (Eds.), *French cultural studies* (pp. 54–96). Oxford. Oxford University Press.

Seman, M. (2010). How a music scene functioned as a tool for urban redevelopment: A case study of Omaha's slowdown project. *City, Culture and Society, 1*(4), 207–215.

Servan-Schreiber, J. J. (1969). *The spirit of May.* New York: McGraw-Hill.

Shernoff, A. (1975). "Master Race Rock" (The Dictators). On *Go Girl Crazy!* [LP]. New York: Sire Records.

Shortridge, J. R. (1991). The concept of the place-defining novel in American popular culture. *The Professional Geographer, 43*(3), 280–291.

Simpson, G. E. (1955). The Ras Tafari movement in Jamaica: A study of race and class conflict. *Social Forces, 34*(2), 167–171. https://doi.org/10.2307/2572834.

Smith, N. (1996). *The new urban frontier: Gentrification and the revanchist city.* London: Routledge.

Smith, H. (2016a, May 9). John Stabb, punk rock headliner of D.C. music scene, dies at 54. *The Washington Post.* Retrieved from https://www.washingtonpost.com/local/if-these-walls-could-talk-theyd-probably-scream/2016/08/01/86bbed62-5751-11e6-831d-0324760ca856_story.html?utm_term=.26f1de1fea31

Smith, T. L. (2016b, December 9). Rock & Roll Hall of Fame 2017: Hard to ignore Bad Brains' influence. Cleveland.com. Retrieved from http://www.cleveland.com/entertainment/index.ssf/2016/12/rock_roll_hall_of_fame_2017_ha.html

Sonnichsen, T. (2013). *Emotion, place, and record collecting in Los Angeles: A postmodernist interpretation.* MA thesis, California State University, Long Beach.

Sonnichsen, T. (2016). Emotional landscapes and the evolution of vinyl record retail: A case study of Highland Park, Los Angeles. In B. Hracs, M. Seman, & T. E. Virani (Eds.), *The production and consumption of music in the digital age* (pp. 190–205). New York: Routledge.

Sparks, L. (2015). A sense of place. *Context: Institute of Historic Building Conservation, 135,* 16–17.

Spracklen, K. (2014). There is (almost) no alternative: The slow 'heat death' of music subcultures and the instrumentalization of contemporary leisure. *Annals of Leisure Research, 17*(3), 252–266.

Stokes, M. (Ed.). (1994). *Ethnicity, identity and music: The musical construction of place.* Oxford: Berg.

Straw, W. (1991). Systems of articulation, logics of change: Communities and scenes in popular music. *Cultural Studies, 5*(3), 368–388.

Straw, W. (2010). *Cities of the night, cultures of the night.* Talk given at the 'My City's Still Breathing: A symposium exploring the arts, artists and the city' conference. Winnipeg, MB.

Straw, W. (2010a). Spectacles of waste. In A. Boutros & W. Straw (Eds.), *Circulation and the city: Essays on urban culture* (pp. 193–213). Montreal: McGill-Queen's University Press.

Straw, W. (2010b). *Cities of the night, cultures of the night.* Paper presented at the My City's still breathing: A symposium exploring the arts, artists and the city, Winnipeg, MB. Viewable at https://vimeo.com/20271240

Straw, W. (2015). Above and below ground. In P. Guerra & T. Moreira (Eds.), *Keep it simple, make it fast: An approach to underground music scenes* (Vol. 1, pp. 403–410). Porto: Universidade do Porto – Faculdade de Letras.

Taylor, H. (2001). *Circling Dixie: Contemporary southern culture through a trans-atlantic lens*. New Brunswick: Rutgers University Press.

Taylor, T. D. (2002). Music and the rise of radio in 1920s America: Technological imperialism, socialization, and the transformation of intimacy. *Historical Journal of Film, Radio and Television, 22*(4), 425–443.

Thien, D. (2009). Feminist methodologies. In R. Kitchin & N. Thrift (Eds.), *International encyclopedia of human geography* (Vol. 4, pp. 71–78). Oxford: Elsevier.

Thien, D. (2011). Emotional life. In V. J. D. Casino, M. Thomas, P. Cloke, & R. Panelli (Eds.), *A companion to social geography* (pp. 309–325). London: Wiley-Blackwell.

Thompson, M. (1979). *Rubbish theory: The creation and destruction of value*. Oxford: Oxford University Press.

Thornton, S. (1996). *Club cultures: Music, media, and subcultural capital*. Middletown: Wesleyan University Press.

Torres, R. (2002). Cancun's tourism development from a Fordist spectrum of analysis. *Tourist Studies, 2*(1), 87–116.

Triggs, T. (2006). Scissors and glue: Punk fanzines and the creation of a DIY aesthetic. *Journal of Design History, 19*(1), 69–83.

Tuan, Y. (1977). *Space and place: The perspective of experience* (10th ed.). Rochester: University of Minnesota Press.

Turrini, J. M. (2013). "Well I don't care about history": Oral history and the making of collective memory in punk rock. *Notes, 70*(1), 59–77.

Urry, J. (1990). *The tourist gaze: Leisure and travel in contemporary societies*. London: Sage.

Vaher, B. (2008). Identity politics reco(r)ded: Vinyl hunters as exotes in time. *Trames, 12*(3), 342–354.

Vecchione, M. A. (Director). (2008). *ANTIFA: Chasseurs de skins*. Documentary. S. Brucher & M. A. Vecchione (Producers). Paris.

Waitt, G. (2010). Doing Foucauldian discourse analysis – Revealing social identities. In I. P. Hay (Ed.), *Qualitative research methods in human geography* (pp. 217–240). Don Mills: Oxford University Press Canada.

Wallach, J. (2002). Exploring class, nation, and xenocentrism in Indonesian cassette retail outlets. *Indonesia, 74*, 79–102.

Wallach, J. (2008). Living the punk lifestyle in Jakarta. *Ethnomusicology, 52*(1), 98–116.

Wallach, J., Berger, H. M., & Greene, P. D. (2011). *Metal rules the globe: Heavy metal music around the world*. Durham: Duke University Press.

Warne, C. (2013). Graphical terrorism? Bazooka, punk and leftist politics at Libération newspaper in 1970s France. *History Workshop Journal, 76,* 212–234.

Watson, A., & Till, K. E. (2010). Ethnography and participant observation. In D. DeLyser, S. Herbert, S. Aitken, M. Crang, & L. McDowell (Eds.), *The Sage handbook of qualitative geography* (pp. 121–137). London: Sage.

Weber, H. (2014). Stereo City: Mobile listening in the 1980s. In M. Gandy & B. Nilsen (Eds.), *The acoustic city* (pp. 156–163). London: Jovis.

Wedren, C. (2016). Skype interview, 18 May.

Weinstein, D. (2011). The globalization of metal. In J. Wallach, H. M. Berger, & P. D. Greene (Eds.), *Metal rules the globe: Heavy metal music around the world* (pp. 34–59). Durham: Duke University Press.

Wetzel, R. (2012). *The globalization of music in history.* London: Routledge.

White, M. (2015). *Popkiss: The life and afterlife of Sarah records.* New York: Bloomsbury.

Wilson, R. E., Gosling, S. D., & Graham, L. T. (2012). A review of Facebook research in the social sciences. *Perspectives on Psychological Science, 7*(3), 203–220.

Wissmann, T. (2014). *Geographies of urban sound.* London: Ashgate.

Wolff, J. (1994). The artist and the flaneur: Rodin, Rilke and Gwen John in Paris. In K. Tester (Ed.), *The Flâneur* (pp. 111–137). London: Routledge.

Wood, N., Duffy, M., & Smith, S. J. (2007). The art of doing (geographies of) music. *Environment and Planning D: Society and Space, 25,* 867–889.

Worley, M. (2014). 'Hey little rich boy, take a good look at me': Punk, class and British Oi!. *Punk & Post Punk, 3*(1), 5–20.

Yow, V. (2010). 'Do I like them too much?' Effects of the oral history interview on the interviewer and vice-versa. In R. Perks & A. Thompson (Eds.), *The oral history reader* (2nd ed., pp. 54–72). London: Routledge.

Zukin, S. (2013). Whose culture? Whose city? In J. Lin & C. Mele (Eds.), *The urban sociology reader* (2nd ed., pp. 349–357). London: Routledge.

The following zines and alternative publications were cited in this book, in alphabetical order. This is hardly a comprehensive list of all the zines I pulled information from or encountered between Paris and DC in this study.

Alien #1, 1994, Grignon (FR)
Altruzine #2, 2001 (DC)
Brand New Age #1, 1983 (DC)
Capitol Crisis #1, 1980 (DC)

Capitol Crisis #2, 1980 (DC)
Capitol Crisis #3, 1981 (DC)
Capitol Crisis #4, 1981 (DC)
Capitol Crisis #5, 1981 (DC)
Comet, Vol. 2 #1, 2001 (DC)
Crack DC #3, 1991 (DC)
DCene #1, 1983 (DC)
Descenes, 1979 (DC)
Dischords, March 1981 (DC)
If This Goes On, 1982 (DC)
Lime Lizard, 1991 (UK)
Maximumrocknroll #91, 1990 (San Francisco)
Maximumrocknroll #93, 1992 (San Francisco)
MilkShake #4, 2010 (Antwerp, Belgium)
Mole #3, 1993 (DC)
Mole #6, 1994 (DC)
Pit #2, 1987 (DC)
Positive Rage #8, 1998 (Paris)
Positive Rage, #9, 2000 (Paris)
Rad Party, various (Paris)
Sidekick #1, 1990 (DC)
The Skills of Defensive Driving #9, 1994 (DC)
Slickzine, 1985 (DC)
Suburban Voice #29, 1990 (Boston)
Truly Needy #10, 1985 (DC)
Trust, 1988 (Germany)
Uno Mas #3, 1991 (DC)
WDC Period #6, 1984 (DC)
Whack, Oct. 1991 (DC)
Whack, Dec. 1991 (DC)
Zone V #1, 1983 (DC)

Index

© The Author(s) 2019
T. Sonnichsen, *Capitals of Punk*, https://doi.org/10.1007/978-981-13-5968-2

The manufacturer's authorised representative in the EU is Springer
Nature Customer Service Centre GmbH, Europaplatz 3, 69115 Heidelberg,
Germany. If you have any concerns regarding our products, please
contact ProductSafety@springernature.com

Printed and bound by CPI Group (UK) Ltd, Croydon, CR0 4YY
29/04/2026
02099458-0002